David Webb · Patrick Vallance (Eds.)

Endothelial Function in Hypertension

Springer
Berlin
Heidelberg
New York
Barcelona
Budapest
Hong Kong
London
Milan
Paris
Santa Clara
Singapore
Tokyo

David Webb · Patrick Vallance (Eds.)

Endothelial Function in Hypertension

With 45 Figures

 Springer

Professor David Webb
Western General Hospital
Clinical Pharmacology Unit
and Research Centre
Edinburgh EH4 2XU
United Kingdom

Professor Patrick Vallance
University College London
The Cruciform Project
London W1P 9LN
United Kingdom

Library of Congress Cataloging-in-Publication Data applied for

```
Endothelial function in hypertension / D. Webb, P. Vallance (eds.).
      p.    cm.
    Includes bibliographical references and index.
    ISBN 3-540-62943-2 (hardcover : alk. paper)
    1. Hypertension--Pathophysiology.  2. Vascular endothelium.
 3. Cardiovascular system--Pathophysiology.  4. Endothelins-
 -Physiological effect.  5. Nitric oxide--Physiological effect.
 I. Webb, David J. (David John), 1953-  .  II. Vallance, P.
 (Patrick), 1960-  .
    [DNLM: 1. Endothelins--physiology.  2. Endothelins--pharmacology-
 -congresses.  3. Endothelium--physiology--congresses.  4. Nitric
 Oxide--physiology--congresses.  5. Hypertension--physiopathology-
 -congresses.    QU 68 E563 1997]
 RC685.H8E55  1997
 616.1'3207--dc21
 DNLM/DLC
 for Library of Congress                              97-17213
                                                          CIP
```

ISBN 3-540-62943-2 Springer-Verlag Berlin Heidelberg New York

Production: PRO EDIT GmbH, D-69126 Heidelberg
Typesetting: STORCH GmbH, D-97353 Wiesentheid
Cover Design: design & production GmbH, D-69121 Heidelberg

SPIN: 10551508   9/3134-5 4 3 2 1 0 – Printed on acid-free paper

Preface

In 1980, Furchgott and Zawadzki demonstrated that the ability of a blood vessel to relax to acetylcholine is entirely dependent upon the presence of an intact endothelium [1]. This startlingly simple and elegant observation heralded a decade of basic research into endothelium-derived vasoactive factors (EDRF). Initially, interest focused on Furchgott's endothelium-derived relaxing factor, which was later identified as the extremely simple molecule, nitric oxide [2]. However, bioassay systems suggested the presence of other dilator and constrictor factors. Many of these additional putative factors remain elusive but one constrictor factor was revealed to the world in an impressively complete piece of science published in 1988. Yanagisawa, Masaki and colleagues [3] demonstrated the synthesis of a twenty one amino acid peptide which accounted for the constrictor activity found in the supernatant of endothelial cells. They identified its structure, synthesised the peptide, showed its biological activity in vitro and in vivo, proposed mechanisms of action and cloned and sequenced the gene. They named the factor endothelin.

The identification of nitric oxide and endothelin has also rekindled interest in established endothelium-derived mediators, including the prostanoids and other arachidonic acid metabolites, and has led to exploration of the possible roles of oxygen radicals such as superoxide. The vascular endothelium has become a major focus for research into disease states or cardiovascular risk factors associated with abnormal vascular tone or reactivity, alterations in cellular adhesion to the vessel wall, increased smooth muscle cell growth and the chronic process of atherogenesis. Endothelial cells also line the chambers of the heart and the same factors released by endocardial endothelium have been implicated in the regulation of myocardial function.

The 1980s was a decade dominated by basic research into endothelial cell biology. The 1990s have seen an explosion of interest in clinical research into the vascular endothelium and the emergence of novel agents with potential therapeutic utility. Thus the 1996 meeting of the ISH was an ideal time to draw together research on constrictor and dilator factors and link the basic discoveries to clinical science and therapeutics. This book contains contributions from basic and clinical scientists who have made substantial advances in our understanding of the physiology, pathophysiology and pharmacology of endothelial vasoactive factors. It should provide a comprehensive and up-to-date overview of a field which is likely to make a significant impact upon the future management of cardiovascular disease.

Spring, 1997   C. R. W. Edwards, P. Vallance, and D. J. Webb

References

1. Furchgott RF, Zawadski JV (1980) The obligatory role of endothelial cells in the relaxation of arterial smooth muscle by acetylcholine. Nature 228:373–376
2. Palmer RMJ, Ferrige AG, Moncada S (1987) Nitric oxide release accounts for the biological activity of endothelium derived relaxing factor. Nature 327:524–526
3. Yanagisawa M, Kurihara H, Kimura S, Tomobe Y, Kobayashi M, Mitsui Y, Yazaki Y, Goto K, Masaki T (1988) A novel potent vasoconstrictor peptide produced by vascular endothelial cells. Nature 332:411–415

Contents

List of Contributors

Daniel D. Borgeson, M.D.
Cardiovascular Diseases and Internal Medicine
Mayo Clinic
200 1st Street SW
Rochester, MN 55905
USA

John C. Burnett, Jr., M.D.
Cardiovascular Diseases and Internal Medicine
Mayo Clinic
200 1st Street SW
Rochester, MN 55905
USA

Richard A. Cohen, M.D.
Vascular Biology Unit
R408 Boston University School of Medicine
80E. Concord St.
Boston, MA 02118
USA

John P. Cooke, M.D., Ph.D.
Stanford University School of Medicine
Falk Cardiovascular Research Center
300 Pasteur Drive
Stanford, CA 94305-5246
USA

Gillian A Gray, PhD
Department of Pharmacology
The University of Edinburgh
1 George Square
Edinburgh, EH8 9JZ
UK

Thomas F. Lüscher, M.D.
Professor and Head of Cardiology
University Hospital
8091 Zürich
Switzerland

Pauline E. McEwan, PhD
Department of Pharmacology
The University of Edinburgh
1 George Square
Edinburgh, EH8 9JZ
UK

Linda, J. McKinley, M.D.
Cardiovascular Diseases
and Internal Medicine
Mayo Clinic
200 1st Street SW
Rochester, MN 55905
USA

Emma J. Mickley, BSc
Clinical Pharmacology Unit
& Research Centre
University of Edinburgh
Western General Hospital
Edinburgh EH4 2XU
UK

Salvador Moncada, M.D.
The Cruciform Project
University College London
140 Tottenham Court Road
London W1P 9LN
UK

Eduardo Nava, M.D.
Department of Physiology
University of Murcia School of Medicine
Murcia
Spain

Patrick J. Pagano
Vascular Biology Unit
R408 Boston University School of Medicine
80E. Concord St.
Boston, MA 02118
USA

F. Javier Salazar, M.D.
Department of Physiology
University of Murcia School of Medicine
Murcia
Spain

Ernesto L. Schiffrin, M.D., Ph.D.
Clinical Research Institute of Montreal
110, Pine avenue west
Montreal, Quebec
H2W 1R7
Canada

Ajay M Shah, M.D.
Department of Cardiology
University of Wales College of Medicine
Heath Park
Cardiff CF4 4XN
UK

Philip S. Tsao, Ph.D.
Stanford University School of Medicine
Falk Cardiovascular Research Center
300 Pasteur Drive
Stanford, CA 94305-5246
USA

Prof. Patrick Vallance, MD FRCP
The Cruciform Project
University College London
140 Tottenham Court Road
London W1P 9LN
UK

David J Webb, MD FRCP
Clinical Pharmacology Unit
& Research Centre
University of Edinburgh
Western General Hospital
Edinburgh EH4 2XU
UK

Nitric Oxide and Hypertension: Physiology and Pathophysiology

P. VALLANCE and S. MONCADA

The field of nitric oxide (NO) research has progressed rapidly and yielded many unexpected insights into cardiovascular control. In 1980 the phenomenon of endothelium-dependent relaxation and the existence of the endothelium-derived relaxing factor (EDRF) were described [1]. Seven years later EDRF was identified as an inorganic gas, NO [2]. Now it is clear that neurones, smooth muscle cells, cardiac myocytes, white cells and platelets, as well as endothelial cells all have the capacity to affect cardiovascular behaviour through generation of NO. Here we discuss the role of NO in the maintenance of normal blood pressure and flow, and describe the significance of the changes that occur in experimental and clinical hypertension. We have reviewed the role of NO in hypotension elsewhere [3].

Synthesis of Nitric Oxide

Nitric oxide is synthesized from the amino acid L-arginine by the action of NO synthases (NOS; Fig. 1). Three isoforms of NOS have been described – endothelial NOS (eNOS), neuronal NOS (nNOS) and an inducible NOS (iNOS). The gene encoding eNOS is located on chromosome 7, the gene for nNOS is on chromosome 12 and that for iNOS is on chromosome 17 [4].

eNOS and nNOS are expressed constitutively in a variety of cells that play a role in cardiovascular physiology (Table 1). These isoforms are calcium/calmodulin-dependent and synthesize relatively low amounts of NO which acts as an intercellular messenger (Fig. 1). In contrast, the iNOS isoform binds calmodulin tightly so that its activity is functionally independent of the prevailing concentration of calcium [5, 6]. Indeed, once expressed iNOS is fully active and generates large quantities of NO. The NO thus generated may produce exaggerated physiological effects but can also be cytostatic or toxic to cells (see below). iNOS is synthesized de novo in cells exposed to certain inflammatory cytokines including tumour necrosis factor (TNF)α, interleukin (IL)-1β and interferon (IFN)α, and its expression may be induced in virtually every type of cell involved in cardiovascular homeostasis [7–10]. Other cytokines such as IL-4, IL-8, IL-10 and transforming growth factor (TGF)β [11–13] decrease the induction of iNOS so that the overall level of expression of this isoform of NOS is likely to depend on the local balance of cytokines. Most cells do not express iNOS constitutively, although constitutive expression of iNOS has been described in the rat kidney [14] and elsewhere (see [4]).

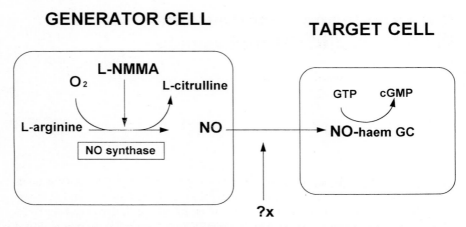

Fig. 1. Nitric oxide (NO) synthase catalyses the synthesis of NO from L-arginine and molecular oxygen. L-citrulline is the byproduct. NO itself might inhibit the activity of NO synthase by interacting with the haem moiety of this enzyme. Physiological effects are produced after NO binds to the haem moiety of guanylate cyclase (GC) and activates this enzyme to produce cyclic guanosine monophosphate (cGMP) from guanosine triphosphate (GTP) in target and generator cells. Molecules (x) that stabilise NO have been proposed. NO is also vulnerable to inactivation by other radicals

Table 1. NOS isoforms in cells affecting cardiovascular homeostasis

	eNOS	nNOS
Endothelium	+	
Neurones		+
Platelets	+	
Neutrophils	+	
Endocardium	+	

eNOS, endothelial nitric oxide synthase; nNOS, neuronal NOS.

Actions of NO

The physiological target for NO is the soluble guanylate cyclase (Fig. 1), a haem-containing heterodimer that synthesizes cyclic GMP (cGMP)[15]. The basal activity of guanylate cyclase is extremely low, but when NO binds to the haem moiety, the activity increases over 400-fold [16]. An increase in the intracellular concentrations of cGMP leads to relaxation of vascular smooth muscle and cardiac myocytes, inhibits platelet aggregation and attenuates white cell and platelet adhesiveness [17]. In certain cells these effects of cGMP are brought about by a decrease in the intracellular concentration of calcium [18].

In addition to effects mediated through cGMP, NO generated in physiological quantities has a reversible inhibitory effect, which is competitive with oxygen, on cytochrome c oxidase [19]. This action might be the basis for the cytostatic effects of NO.

In large quantities, however, NO can produce toxic effects. These are thought to be due to the generation of peroxynitrite ($ONOO^-$), a product of the reaction between NO and superoxide anion (O_2^-). If produced at the mitochondrial level $ONOO^-$ can irreversibly inhibit complexes I-III [20], thus causing cytotoxicity. It may also cause damage by nitrating tyrosine (and other) residues on proteins. However, $ONOO^-$ may be inactivated by combining with a range of molecules such as thiols, low molecular mass antioxidants and sugars. Thus the net effect of peroxynitrite appears to be critically dependent on the local concentration of thiols and other molecules that may act as scavengers [21].

Nitric Oxide and Vasodilatation

Inhibition of endogenous generation of NO with an inhibitor of NOS, such as N^G-monomethyl-L-arginine (L-NMMA), causes vasoconstriction of small arteries and arterioles and elevates resting blood pressure [22, 23]. This effect has been demonstrated in a variety of species including man [24–26]. Indeed, in virtually every arterial bed studied, inhibition of NOS causes approximately 50% reduction in resting blood flow at constant perfusion pressure (Fig. 2), ie a near doubling of vascular resistance [27]. In contrast to the significant basal NO-mediated dilatation in small arteries, basal generation of NO appears to contribute less to the regulation of tone in conduit arteries

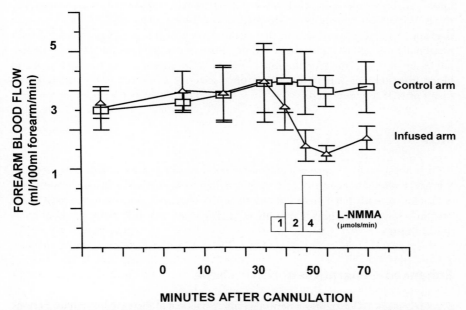

Fig. 2. The effects of L-NMMA on resting forearm blood flow. L-NMMA was infused into one arm and the other arm acted as a control. L-NMMA produced a dose-dependent fall in blood flow as endogenous nitric oxide synthesis was blocked

and is absent in many veins [27]. The reasons for this are not clear, since conduit arteries and veins contain abundant eNOS and are able to generate significant amounts of NO when stimulated by appropriate agonists (see below).

eNOS

It has been assumed that the increase in vascular resistance produced by NOS inhibitors is due largely or exclusively to inhibition of eNOS in vascular endothelium. In support of this assumption, mice lacking the gene encoding eNOS (eNOS knockout mice) are hypertensive [28]. However, these mice show a paradoxical depressor response to NOS inhibitors [28], the reason for which has not been elucidated.

nNOS

Certain vessels are supplied by non-adrenergic, non-cholinergic (NANC) nerves that contain nNOS and release NO onto the adventitial side of blood vessels [29]. Such nitrergic nerves are found in the corpus cavernosum [30, 31] (where they contribute to the process of erection) and in the brain [32] (where they may act to couple neuronal activation to changes in blood flow). nNOS knockout mice are normotensive but do not show the usual changes in cerebrovascular reactivity in response to NOS inhibitors [33]. Experiments in which vasoconstriction of the dog basilar artery in response to a NOS inhibitor is measured in the presence or absence of ganglion blockade suggest that approximately two-thirds of the vasodilatation in this vessel is associated with neuronally-derived NO, with the remaining third due to NO generated by non-neuronal tissue [see 34]. However, the role of nNOS in the regulation of vascular tone and blood pressure will only become clear once selective inhibitors of the different isoforms of NOS are identified.

iNOS

iNOS is not normally expressed in healthy blood vessels and it has been assumed that this isoform does not contribute to the physiological regulation of vascular tone. However, it has recently been observed that the iNOS knockout mouse has a resting blood pressure 10–15 mmHg higher than its wild-type litter mates (D. Rees, personal communication).

Enhanced Generation of Nitric Oxide

Basal release of NO provides a natural counter-balance to the vasoconstrictor actions of the sympathetic nervous system and a mechanism by which the vessel can adjust its tone rapidly in response to local changes in its chemical or physical environment.

The vascular endothelium has receptors for a variety of circulatory and local hormones (including bradykinin, substance P, noradrenaline, adrenaline, serotonin) and has ion channels capable of responding to changes in pressure, stretch or shear stress [27]. Receptor occupation by an agonist, or the opening of certain ion channels, leads to elevation of intracellular calcium and this activates eNOS to generate NO. It seems likely that the shear stress produced by pulsatile flow contributes to the "basal" generation of NO in the arterial system, although the local "chemical" environment may also contribute. Indeed, many small arteries continue to generate NO basally when studied in vitro [35], suggesting that continued shear stress is not necessary for the basal generation of NO. Shear stress-induced generation of NO accounts for the phenomenon of flow-mediated dilatation and might be a mechanism by which vessels keep shear forces constant in the face of changes in flow. Shear stress in vitro and chronic exercise have been shown to enhance eNOS gene expression in the aortic endothelium (see [36]). This has led to the identification of consensus sites on the eNOS gene for regulation by shear stress and other stimuli (see [37]). Interestingly, during pregnancy and treatment with oestradiol, mRNAs for both eNOS and nNOS are also increased [38]. This oestrogen-induced increase in NO synthesis could contribute to the decrease in vascular tone and contractility that occurs during pregnancy, the associated increase in gastrointestinal transit time and the reduced incidence of heart disease in pre-menopausal women.

Animal Models of Hypertension

Endothelium-dependent relaxation is impaired in many models of hypertension [39–42]. In addition, in certain models, the constrictor response to NOS inhibitors is also diminished suggesting that basal NO-mediated dilatation is impaired. However, there are other apparently contradictory reports [43], some even demonstrating that NO generation is increased in hypertension [44]. Much of the apparent contradiction is methodological in origin but some relates to the model of hypertension studied.

In certain types of experimental hypertension there is unequivocal evidence for an impairment in the generation of NO. An example is provided by the Sabra rat [39]. Sabra rats can be divided into two groups, namely hypertension-resistant and hypertension-prone (salt sensitive). Compared to the hypertension-resistant Sabra rats, vasorelaxant responses to acetylcholine are diminished in the hypertension-prone animals, the constrictor responses to L-NMMA are reduced, less eNOS is present in the vasculature, and the circulating levels of nitrite and nitrate (breakdown products of NO) are lower [39]. The activity of the L-arginine: NO pathway also appears to be diminished in the Dahl salt-sensitive rat [42], and salt-resistant animals are rendered salt-sensitive following treatment with NOS inhibitors [45]. It appears, therefore, that certain animal models of salt-sensitive hypertension are associated with deficient NO-mediated dilatation and this deficiency might contribute directly to the salt sensitivity. The mechanisms for such an effect are not yet known. Although changes in eNOS are implicated in the Sabra rat, hypertension in the Dahl rat has been associated with the iNOS gene [46].

In other models the data are less clear cut. Increased generation of NO has been detected in the spontaneously hypertensive rat although it has been proposed that the coincident generation of O_2^- might "neutralise" the NO before it reaches the smooth muscle [43]. Thus in this model, an apparent increase in eNOS activity might be associated with a decrease in the *functional* activity of the pathway, depending on the level of O_2^- generated. The effects of antihypertensive therapy on generation of NO are also inconclusive. Certain data suggest that defective endothelium-dependent dilatation is restored to normal by such treatment [47, 48], whilst other data suggest that such a defect is irreversible [49].

Human Hypertension

The vast majority of studies of the L-arginine: NO pathway in human hypertension have employed the technique of venous occlusion plethysmography to study the forearm arterial bed. Investigators have used infusions of acetylcholine to stimulate NO-mediated vasodilatation, or have used L-NMMA to define the contribution made by NO to the resting vascular tone

Since the initial report [50], in excess of 20 studies have documented reduced responses to acetylcholine in patients with primary (essential) or secondary hypertension, and even in the offspring of hypertensive patients [51]. However, one large well controlled study [52], and at least one other smaller study [53], failed to detect any difference in responsiveness to acetylcholine in normotensive and hypertensive subjects. The data with acetylcholine are difficult to interpret with confidence since the response to this agonist is extremely variable and is not mediated solely by NO [24, 54]. Furthermore, acetylcholine is rapidly destroyed by cholinesterase so that nearly 100% disappears during flow down the forearm from the brachial artery to the wrist [55]. In addition, testing the ability of the system to increase its activity in response to an agonist is of uncertain physiological significance.

Studies with NOS inhibitors appear to give more information on basal NO. Several reports show that the constrictor response to L-NMMA is diminished in patients with essential hypertension [56–61]. Reduction of the blood pressure with antihypertensive drugs such as angiotensin converting enzyme (ACE) inhibitors and calcium channel blockers appears to restore the response to L-NMMA towards normal [60, 61], suggesting that any "defect" is readily reversible and might be a response to the high pressure. Interestingly, platelets of patients with essential hypertension are also less sensitive to L-NMMA [61] compared with platelets from normotensive subjects, indicating that changes in the pathway might not be limited to the vasculature in this condition.

Studies in the forearm suggest that dilatation due to basal release of NO is less in hypertensive than in normotensive subjects. This does not necessarily mean that there is a defect in NOS enzymes, or that less NO is being generated. Alternative explanations include the possibility that more NO is destroyed before it reaches the smooth muscle, that the smooth muscle is less responsive to NO, or even that changes in the cellular uptake or metabolism of L-NMMA occur in hypertension. However, evidence in favour of diminished NO generation in human hypertension comes from a study using ^{15}N-arginine which demonstrated reduced generation of ^{15}N nitrate over a 24-h time period [62]. Similar to the response to L-NMMA, the generation of ^{15}N nitrate correlates inversely with blood pressure, suggesting that the higher the pressure the lower the rates of NO generation [62]. Further studies will be required to test this hypothesis directly and to identify a mechanism by which NO-mediated dilatation may be affected by pressure. Polymorphisms identified in the eNOS gene have not been associated with altered blood pressure [63, 64].

Overall, the data in humans suggest that the functional activity of the L-arginine: NO pathway is diminished in essential hypertension and that the change is secondary to the raised pressure rather than causative.

Consequences of Diminished Nitric Oxide

Reduction in NO synthesis leads to arterial vasoconstriction and hypertension. Chronic administration of NOS inhibitors causes sustained hypertension [65] and produces renal damage similar to that seen in malignant hypertension [66]. Loss of NO-mediated dilatation also appears to "stiffen" the cardiovascular system, leading to decreased vascular compliance and a widened pulse pressure, and diminishes the ability to accommodate to rapid changes in intravascular volume without increasing arterial pressure [67, 68]. NO plays a role in facilitating sodium excretion so that systemic inhibition of NOS promotes salt and water retention [69, 70]. Together these findings suggest that a reduction in NO-mediated dilatation will increase arterial resistance and enhance the susceptibility of the cardiovascular system to pressor stimuli.

In addition to effects on vascular tone and pressure, defects in the L-arginine: NO pathway enhance experimental atherogenesis (see chapter by Cooke, pages 29–38 this volume). Thus the finding that the basal functional activity of the pathway appears to decrease with increasing pressure provides a possible mechanism to link increased arterial pressure to atherogenesis. Furthermore, impairment of NO-mediated dilatation has also been described in diabetes, hypercholesterolaemia and smokers, suggesting that alterations in NO generation or effect provide one mechanism by which different risk factors may promote the common end point of atherogenesis. In support of a role for altered NO generation in atherogenesis, a rare polymorphism of the eNOS gene appears to be associated with the presence of severe atherosclerosis [71].

Therapeutics

ACE inhibitors prevent the breakdown of bradykinin, a mediator that stimulates the endothelium to generate NO [27]. Whether this contributes to any of the beneficial effects of ACE inhibitors remains to be determined. Existing NO-donors (e.g. glyceryl trinitrate) act predominantly by dilating vessels which do not generate endogenous NO basally (ie veins). However, NO-donors with a more balanced arterio-venous profile [72] are available for experimental use and some of these also have potent anti-platelet effects [73, 74] and prevent the adhesion of white cells to the endothelium.

Finally, L-arginine appears to have beneficial effects on vascular reactivity in patients with hypercholesterolaemia [75,76] and might lower blood pressure under certain conditions, although the mechanism of these effects is far from clear [77]. The discovery of the L-arginine: NO pathway has radically changed our understanding of the physiological regulation of blood pressure, and has led directly to the development of novel models of hypertension and atherosclerosis, which are giving exciting insights into diseases of the cardiovascular system. These developments will no doubt lead in the future to novel therapies for the treatment and prevention of cardiovascular disease.

References

1. Furchgott RF, Zawadzki JV (1980) The obligatory role of endothelial cells in the relaxation of arterial smooth muscle by acetylcholine. Nature 288:373–376
2. Palmer RMJ, Ferrige, AG, Moncada S (1987) Nitric oxide release accounts for the biological activity of endothelium-derived relaxing factor. Nature 327:524–526
3. Vallance P, Moncada S (1993) Role of endogenous nitric oxide in septic shock. New Horizons 1:77–86
4. Moncada S, Higgs A (1995) Molecular mechanisms and therapeutic strategies related to nitric oxide. FASEB J 9:1319–1330
5. Bogle R, Vallance P (1996) Functional effects of econazole on inducible nitric oxide synthase: production of a calmodulin-dependent enzyme. Br J Pharmacol 117:1053–1058
6. Cho HJ, Xie Q-W, Calaycay J, Mumford RA, Swiderek KM, Lee TD, Nathan CF (1992) Calmodulin is a subunit of nitric oxide synthase from macrophages. J Exp Med 176:599–604
7. Bhagat K, Vallance P (1996) Inducible nitric oxide synthase in the cardiovascular system. Heart 75:218–220
8. Cook TH, Bune AJ, Taylor M, Loi RK, Cattell V (1994) Cellular localisation of inducible nitric oxide in endotoxic shock in the rat. Clin Sci 87 179–186
9. de Belder A, Radomski M, Hancock V, Brown A, Moncada S, Martin J (1995) Megakaryocytes from patients with coronary atherosclerosis express the inducible nitric oxide synthase. Arterioscler Thromb Vasc Biol 15:637–641
10. Radomski MW, Vallance P, Whitley G, Foxwell N, Moncada S (1993) Platelet adhesion to human vascular endothelium is modulated by constitutive and cytokine induced nitric oxide. Cardiovasc Res 89:2070–2078
11. McCall TB, Palmer RM, Moncada S (1992) Interleukin-8 inhibits the induction of nitric oxide synthase in rat peritoneal neutrophils. Biochem Biophys Res Commun 186:680–685
12. Oswald IP, Gazzinelli RT, Sher A, James SL (1992) IL-10 synergizes with IL-4 and transforming growth factor beta to inhibit macrophage cytotoxic activity. J Immunol 148:3578–3582
13. Liew FY (1994) Regulation of nitric oxide synthesis in infectious and autoimmune diseases. Immunol Lett 43:95–98
14. Mohaupt MG, Elzie JI, Ahn KY, Clapp WL, Wilcox CS, Kone BC (1994) Differential expression and induction of mRNAs encoding two inducible nitric oxide synthases in rat kidney. Kidney Int 46:653–665
15. McDonald LJ, Murad F (1996) Nitric oxide and cyclic GMP signaling. Proc Soc Exp Biol Med 211:1–6
16. Stone JR, Sands RH, Dunham WR, Marletta MA (1996) Spectral and ligand-binding properties of an unusual hemoprotein, the ferric form of soluble guanylate cyclase. Biochemistry 35:3258–3262
17. Calver A, Collier J, Vallance P (1993) Nitric oxide and the control of human vascular tone in health and disease. Eur J Med 2:48–53
18. Kai H, Kanaide H, Matsumoto T, Nakamura M (1987) 8-Bromoguanosine 3': 5'-cyclic monophosphate decreases intracellular free calcium concentrations in cultured vascular smooth muscle cells from rat aorta. FEBS Lett 221:284–288
19. Cleeter MWJ, Cooper JM, Darley-Usmar VM, Moncada S, Schapira AHV (1994) Reversible inhibition of cytochrome c oxidase, the terminal enzyme of the mitochondrial respiratory chain by nitric oxide. FEBS Lett 345:50–54

20. Lizasoain I, Moro MA, Knowles RG, Darley-Usmar V, Moncada S (1996) Nitric oxide and peroxynitrite exert distinct effects on mitochondrial respiration which are differentially blocked by glutathione or glucose. Biochem J 314:877–880
21. Villa LM, Salas E, Darley-Usmar VM, Radomski MW, Moncada S (1994) Peroxynitrite induces both vasodilatation and impaired vascular relaxation in the isolated perfused rat heart. Proc Natl Acad Sci USA 91:12383–12387
22. Rees DD, Palmer RM, Moncada S (1989) Role of endothelium-derived nitric oxide in the regulation of blood pressure. Proc Natl Acad Sci USA 86:3375–3378
23. Aisaka K, Gross SS, Griffith OW, Levi R (1989) N^G-methylarginine, an inhibitor of endothelium-derived nitric oxide synthesis, is a potent pressor agent in the guinea pig: does nitric oxide regulate blood pressure in vivo? Biochem Biophys Res Commun 160:881–886
24. Vallance P, Collier J, Moncada S (1989) Effects of endothelium-derived nitric oxide on peripheral arteriolar tone in man. Lancet 2:997–1000
25. Haynes WG, Noon JP, Walker BR, Webb DJ (1993) Inhibition of nitric oxide synthesis increases blood pressure in healthy humans. J Hypertens 11:1375–1380
26. Stamler JS, Loh E, Roddy MA, Currie KE, Creager MA (1994) Nitric oxide regulates basal systemic and pulmonary vascular resistance in healthy humans. Circulation 89:2035–2040
27. Calver A, Collier J, Vallance P (1993) Nitric oxide and cardiovascular control. Exp.Physiol 78:303–326
28. Huang PL, Huang Z, Mashimo H, Bloch KD, Moskowitz MA, Bevan JA, Fishman MC (1995) Hypertension in mice lacking the gene for endothelial nitric oxide synthase. Nature 377:239–242
29. Toda N (1995) Regulation of blood pressure by nitroxidergic nerve. J Diabetes Complications 9:200–202
30. Ignarro LJ, Bush PA, Buga GM, Wood KS, Fukuto JM, Rajfer J (1990) Nitric oxide and cyclic GMP formation upon electrical field stimulation cause relaxation of corpus cavernosum smooth muscle. Biochem Biophys Res Commun. 170:843–850
31. Leone AM, Wiklund NP, Hokfelt T, Brundin L, Moncada S (1994) Release of nitric oxide by nerve stimulation in the human urogenital tract. Neuroreport 5:733–736
32. Faraci FM, Brian JE Jr (1994) Nitric oxide and the cerebral circulation. Stroke 25:692–703
33. Irikura K, Huang PL, Ma J, Lee DW, Dalkara T, Fishman MC, Dawson TM, Snyder SH, Moskowitz MA (1995) Cerebrovascular alterations in mice lacking neuronal nitric oxide synthase gene expression. Proc Natl Acad Sci USA 92:6823–6827
34. Toda N, Okamura T (1996) Nitroxidergic nerve: regulation of vascular tone and blood flow in the brain. Journal of Hypertension 14:423–434
35. Luscher TF, Dohi Y, Tschudi M (1992) Endothelium-dependent regulation of resistance arteries: alterations with aging and hypertension. J Cardiovasc Pharmacol 19:S34–S42
36. Sessa WC (1994) The nitric oxide synthase family of proteins. J.Vasc.Res. 31;131–143
37. Nathan C, Xie Q-W (1994) Regulation of biosynthesis of nitric oxide. J.Biol.Chem. 269;13725–13728
38. Weiner CP, Lizasoain I, Baylis SA, Knowles RG, Charles IC, Moncada S (1994) Induction of calcium-dependent nitric oxide synthases by sex hormones. Proc. Natl. Acad. Sci. USA 91:5212–5216
39. Rees D. Ben-Ishay D, Moncada S (1996) Nitric oxide and the regulation of blood pressure in the hypertension-prone and hypertension-resistant Sabra rat. Hypertension 28:367–371
40. Dohi Y, Thiel MA, Buhler FR, Luscher TF (1990) Activation of endothelial L-arginine pathway in resistance arteries. Effect of age and hypertension. Hypertension 16:170–179
41. Baylis C, Vallance P (1996) Nitric oxide and blood pressure: effects of nitric oxide deficiency. Current Opinion in Neph 5:80–88
42. Goegehold MA (1992) Reduced influence of nitric oxide on arteriolar tone in hypertensive Dahl rats. Hypertension 19:290–295
43. Tschudi MR, Mesaros S, Luscher TF, Malinski T (1996) Direct in situ measurement of nitric oxide in mesenteric resistance arteries. Increased decomposition by superoxide in hypertension. Hypertension 27:32–35
44. Kelm M, Feelisch M, Krebber T, Motz W, Strauer BE (1992) The role of nitric oxide in the regulation of coronary vascular resistance in arterial hypertension: comparison of normotensive and spontaneously hypertensive rats. J. Cardiovasc Pharmacol 20:S183–S186
45. Tolins JP, Shultz PJ (1994) Endogenous nitric oxide synthesis determines sensitivity to the pressor effect of salt. Kidney Int 46:230–236
46. Deng AY, Rapp JP (1995) Locus for the inducible, but not a constitutive, nitric oxide synthase cosegregates with blood pressure in the Dahl salt-sensitive rat. J Clin Invest 95:2170–2177

47. Tschudi Mr, Criscione L, Novosel D, Pfeiffer K, Luscher TF (1994) Antihypertensive therapy augments endothelium-dependent relaxations in coronary arteries of spontaneously hypertensive rats. Circulation 89:2212–2218

48. Luscher TF, Vanhoutte PM, Raij L (1987) Antihypertensive treatment normalizes decreased endothelium-dependent relaxations in rats with salt-induced hypertension. Hypertension 9:III193–III197

49. Luscher TF (1994) The endothelium in hypertension:bystander, target or mediator? J Hypertens Suppl 12:S105–S116

50. Linder L, Kiowski W, Buhler FR, Luscher TF (1990) Indirect evidence for release of endothelium-derived relaxing factor in human forearm circulation in vivo. Blunted response in essential hypertension. Circulation 81:1762–1767

51. Taddei S, Virdis A, Mattei P, Ghiadoni L, Sudano I, Salvetti A (1996) Defective L-arginine-nitric oxide pathway in offspring of essential hypertensive patients. Circulation 94:1298–1303

52. Cockcroft JR, Chowienszky PJ, Benjamin N, Ritter JM (1994) Preserved endothelium-dependent vasodilatation in patients with essential hypertension. N Engl J Med 330:1036–1040

53. Bruning TA, Chang PC, Hendriks MG, Vermeij P, Pfaffendorf M, van-Zwieten PA (1995) In vivo characterization of muscarinic receptor subtypes that mediate vasodilatation in patients with essential hypertension. Hypertension 26:70–77

54. Chowienczyk PJ, Cockcroft JR, Ritter JM (1995) Inhibition of acetylcholinesterase selectively potentiates N^G-monomethyl-L-arginine-resistant actions of acetylcholine in human forearm vasculature. Clin Sci 88:111–117

55. Duff E, Greenfield ADM, Shepherd JT, Thompson ID (1953) A quantitative study of the response to acetylcholine and histamine of the blood vessels of the human hand and forearm. J Physiol 120:160–170

56. Calver A, Collier J, Moncada S, Vallance P (1992) Effect of local intra-arterial N^G-monomethyl-L-arginine in patients with hypertension: the nitric oxide dilator mechanism appears abnormal. J Hypertens 10:1025–1031

57. Lyons D, Webster J, Benjamin N (1994) The effect of antihypertensive therapy on responsiveness to local intra-arterial N^G-monomethyl-L-arginine in patients with essential hypertension. J Hypertens 12:1047–1052

58. Taddei S, Virdis A, Mattei P, Natali A, Ferrannini E, Salvetti A (1995) Effect of insulin on acetylcholine-induced vasodilatation in normotensive subjects and patients with essential hypertension. Circulation 92:2911–2918

59. Panza JA, Casino PR, Kilcoyne CM, Quyyumi AA (1993) Role of endothelium-derived nitric oxide in the abnormal endothelium-dependent vascular relaxation of patients with essential hypertension. Circulation 87:1468–1474

60. Calver A, Collier J, Vallance P (1994) Forearm blood flow responses to a nitric oxide synthase inhibitor in patients with treated essential hypertension. Cardiovasc Res 28:1720–1725

61. Cadwgan TM, Benjamin N (1993) Evidence for altered platelet nitric oxide synthesis in essential hypertension. J Hypertens 11:417–420

62. Forte P, Copland M, Smith L, Milne E, Sutherland J, Benjamin N (1997) Basal nitric oxide synthesis is reduced in essential hypertension. Lancet 349:837–842

63. Hunt SC, Williams CS, Sharma AM, Inoue I, Williams RR, Lalouel JM (1996) Lack of linkage between the endothelial nitric oxide synthase gene and hypertension. J Hypertens 10:27–30

64. Bonnardeaux A, Nadaud S, Charru A, Jeunemaitre X, Corvol P, Soubrier F (1995) Lack of evidence for linkage of the endothelial cell nitric oxide synthase gene to essential hypertension. Circulation 91:96–102

65. Ribeiro MO, Antunes E, de Nucci G, Lovisolo SM, Zatz R (1992) Chronic inhibition of nitric oxide synthesis. A new model of arterial hypertension. Hypertension 20:298–303

66. Yamada SS, Sassaki AL, Fujihara CK, Malheiros DM, de-Nucci G, Zatz R (1996) Effect of salt intake and inhibitor dose on arterial hypertension and renal injury induced by chronic nitric oxide blockade. Hypertension 27:1165–1172

67. Calver A, Collier J, Green D, Vallance P (1992) Effect of acute plasma volume expansion on peripheral arteriolar tone in healthy subjects. Clin Sci 83:541–547

68. Salas SP, Altermatt F, Campos M, Giacaman A, Rosso P (1995) Effects of long-term nitric oxide synthesis inhibition on plasma volume expansion and fetal growth in the pregnant rat. Hypertension 26:1019–1023

69. Salazar FJ, Alberola A, Pinilla JM, Romero JC, Quesada T (1993) Salt-induced increase in arterial pressure during nitric oxide synthesis inhibition. Hypertension 22:49–55

70. Shultz PJ, Tolins JP (1993) Adaptation to increased dietary salt intake in the rat. Role of endogenous nitric oxide. J Clin Invest 91:642–650

71. Wang-XL, Sim AS, Badenhop RF, McCredie RM, Wilcken DE (1996) A smoking-dependent risk of coronary artery disease associated with a polymorphism of the endothelial nitric oxide synthase gene. Nat Med. 2:41–45
72. MacAllister RJ, Calver AL, Riezebos J, Collier J, Vallance P (1995) Relative potency and arteriovenous selectivity of nitrovasodilators on human blood vessels: an insight into the targeting of nitric oxide delivery. J Pharmacol Exp Ther 273:154–160
73. de Belder AJ, MacAllister R, Radomski MW, Moncada S, Vallance P (1994) Effects of S-nitroso-glutathione in the human forearm circulation: evidence for selective inhibition of platelet activation. Cardiovasc Res 28:691–694
74. Radomski MW, Rees DD, Dutra A, Moncada S (1992) S-nitroso-glutathione inhibits platelet activation in vitro and in vivo. Br J Pharmacol 107:745–749
75. Drexler H, Zeiher AM, Meinzer K, Just H (1991) Correction of endothelial dysfunction in coronary microcirculation of hypercholesterolaemic patients by L-arginine. Lancet 338:1546–1550
76. Clarkson P, Adams MR, Powe AJ, Donald AE, McCredie R, Robinson J, McCarthy SN, Keech A, Celermajer DS, Deanfield JE (1996) Oral L-arginine improves endothelium-dependent dilatation in hypercholesterolemic young adults. J Clin Invest 97:1989–1994
77. Hishikawa K, Nakaki T, Suzuki H, Kato R, Saruta T (1992) L-arginine as an antihypertensive agent. J Cardiovasc Pharmacol 20:S196–197

Endothelial Function in Hypertension: Role of Nitric Oxide

E. Nava, F. J. Salazar and T. F. Lüscher

Nitric oxide (NO) is the endogenous vasodilator produced by the endothelium. Although NO release seems to be influenced by changes in blood pressure, whether its production is or not affected in hypertension has been very controversial. Recent research work is showing the existence of profound differences in the role of NO in hypertension. This role seems to vary depending on the class of hypertension. In spontaneous and renovascular hypertension, the production of NO is increased probably as a compensatory mechanism. However, in genetic hypertension NO is inefficacious, presumably because of increased inactivation by oxidative radicals. In this form of hypertension an increased production of contractile factors or a decreased release of hyperpolarizing factors seems to be involved. In salt-dependent hypertension, NO production is impaired probably due to a deficiency of the substrate for NO synthase. In human hypertension, pharmacological experiments reveal an impaired NO dilator mechanism. In pulmonary hypertension, the use of NO gas inhalation has been proposed as a future therapy for this condition. This chapter reviews the evidence, concentrating on the evidence from animal models.

Physiological Roles of the Endothelium

The endothelium is a monolayer of cells lying on the vascular wall which for years was considered to be only a protective barrier. In the past two decades, however, it has been shown that the endothelium plays indeed an active role in the regulation of vascular smooth muscle cell function and vascular tone. In the mid seventies, Sir John Vane's group discovered prostacyclin, a potent vasodilator and platelet antiaggregator released by endothelial cells [1]. In 1980, Furchgott and Zawadski established the obligatory role endothelial cells play in the relaxation to acetylcholine. The factor released by the endothelium was named endothelium-derived relaxant factor ([2]; EDRF) and was later shown to be the gas NO by the group led by Salvador Moncada [3]. The endothelium is also a source of other factors such as the putative endothelium-derived hyperpolarizing factor [4] and potent vasoconstrictors such as endothelin-1, thromboxane A_2 and prostaglandin H_2 [5].

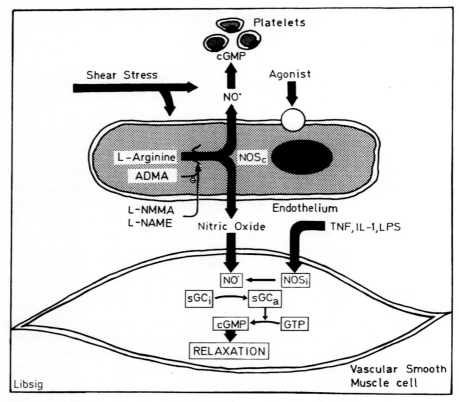

Fig. 1. The L-arginine/NO pathway in the blood vessel wall. *sGC*, soluble guanylate cyclase; *cGMP*, cyclic 3′,5′-guanosine monophosphate; *LPS*, lipopolysaccharide; *TNF*, tumor necrosis factor; *IL-1*, interleukin 1; *cNOS/iNOS*, constitutive/inducible NO synthase; *ADMA*, asymmetrical dimethylarginine; *L-NMMA*, monomethyl-L-arginine; *L-NAME*, nitro-L-arginine methylester

Nitric Oxide in the Cardiovascular System

Since its discovery, an increasing body of evidence shows that NO is a widespread biological mediator implicated in many physiological and pathophysiological processes, including a variety of cardiovascular diseases [6]. Nitric oxide is synthesized from L-arginine by a family of enzymes called NO synthases (NOS) (Fig. 1) [7]. One isoform of these is a calcium-dependent enzyme which is present in endothelial cells and platelets [7]. This enzyme, eNOS, is constitutively expressed in these cells and is modulated by shear stress on the endothelium and by a variety of receptor-activated agonists as well as hormones [8–10]. Nitric oxide synthesized by eNOS plays a crucial role in the regulation or blood pressure, vascular tone and platelet aggregation [11–13]. Another type of NOS is calcium-independent and inducible by immunological stimuli (iNOS;

Fig. 1) [7]. Within the cardiovascular system iNOS is present in endothelial and smooth muscle cells as well as macrophages. This enzyme takes part in several immuno-pathological processes [7].

Nitric Oxide and Hypertension

Nitric oxide may be involved in the pathogenesis of hypertension as well as its complications. Inhibitors of NO production cause endothelium-dependent contractions of isolated arteries, decrease blood flow and induce pronounced and sustained hypertension when infused intravenously or given orally in vivo [11, 12, 14]. The cardiovascular system is exposed to a continuous NO-dependent vasodilatory tone [11] and withdrawal of it mimics many features of human hypertension including target organ damage [15, 16]. Of particular interest in this context is the fact that NO not only acts as a vasodilator, but also inhibits platelet adhesion and aggregation [13] as well as migration and proliferation of vascular smooth muscle cells [17].

In situations of NO deprivation, which can be simulated by treating animals chronically with NO synthesis inhibitors, the endothelium can replace the functional abilities of NO provided antihypertensive treatment is supplied (Fig. 2) [14]. This is probably because the endothelium increases the release of alternative relaxing factors which in normal conditions are not present.

The role of the endothelium and NO in systemic hypertension is very controversial. Although an impaired release of relaxing factors may partly underlie the pathogenesis of hypertension [18], it now appears that endothelium-dependent relaxation is heterogeneously affected in this condition. In some vascular beds of hypertensive rats such as the aorta, mesenteric, carotid and cerebral vessels, endothelium-dependent relaxation is impaired [18–20]. In contrast, in coronary and renal arteries of spontaneously hypertensive rats, endothelial function does not seem to be affected by high blood pressure [20, 21].

Modulation of Nitrix Oxide Release by Changes in Blood Pressure

Studies performed in our lab demonstrate that acute, pharmacologically-induced, drops in blood pressure cause a decreased release of NO (Fig. 3) [22] and, on the contrary, elevations in pressure are followed by an increased production [23], suggesting that high blood pressure upregulates NO production and vice versa. Recent observations demonstrate that an induction (i.e., de novo synthesis of protein) of endothelial NOS takes place in hypertension [24]. The mechanism by which high blood pressure leads to an increased production of NO is not clear yet. It is known that the release of NO by endothelial cells can be altered by changes in shear stress [25] and that mRNA and protein for eNOS can be induced by mechanical forces different from shear forces [26]. It is plausible that not only shear stress, but also other mechanical factors such as blood pressure and pulsatile stretch contribute to this phenomenon.

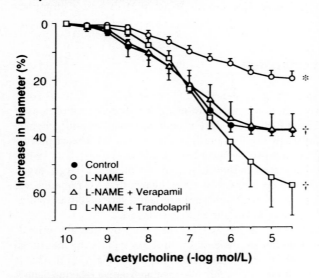

a) Chronic Treatment

Increase in Diameter (%)

- ● Control
- ○ L-NAME
- △ L-NAME + Verapamil
- □ L-NAME + Trandolapril

Acetylcholine (-log mol/L)

Fig. 2a,b. Concentration-response curves to acetylcholine in mesenteric arteries treated with placebo *(control)*, N^wnitro-L-arginine methyl ester *(L-NAME)*, LNAME plus verapamil and L-NAME plus trandolapril. Results are expressed as percent of the increase in the intraluminal vascular diameter from the precontracted condition by norepinephrine (2×10^{-6} mol/l). *$p < 0.05$ vs. control group; $p < 0.05$ vs. L-NAME group (ANOVA + Bonferroni). (Modified from [14])

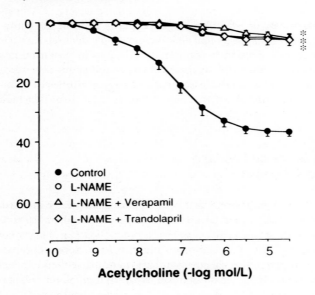

b) Acute Treatment

- ● Control
- ○ L-NAME
- △ L-NAME + Verapamil
- ◇ L-NAME + Trandolapril

Acetylcholine (-log mol/L)

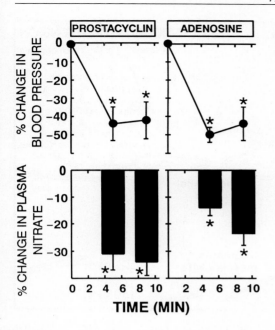

Fig. 3. Effects of lowering blood pressure with a continuous infusion of prostacyclin or adenosine on the plasma concentration of nitrate. The decrease in plasma nitrate suggests a downregulation of NO production when blood pressure diminishes. (From [22])

Nitric Oxide in Spontaneous Hypertension

The spontaneously hypertensive rat (SHR) is a genetic model of hypertension which we have investigated for many years. In the late eighties it was demonstrated that the impaired relaxations of aortic segments from SHR are not caused by a deterred production of endothelium-derived relaxing factor (EDRF) but instead, by an increased release of prostanoid contractile factors (Fig. 4) [18, 27]. With the advent of the NO age and using new methodologies we have confirmed the initial observations on the release of EDRF in this form of hypertension. The activity of eNOS is higher in the heart, kidney and mesenteric resistance arteries obtained from SHR compared to age-matched normotensive rats [24, 28, 29]. Moreover, the concentration of the oxidative product of NO, nitrate, is higher in hypertensive rats as compared to normotensive controls [24]. In contrast, prehypertensive young SHR exhibit similar nitrate levels as the normotensive. These observations demonstrate that the basal release of NO is increased in rats with spontaneous hypertension and that this increased production is directly related with the high blood pressure of the animals. Interestingly, Hirata's group has found that, although the vasoactive effects of acetylcholine in the kidney are actually abnormal, the measured release of NO from renal vessels is not diminished but slightly higher in the SHR [30]. Treatment with inhibitors of the endothelium-derived hyperpolarizing factor (EDHF) bring the vasoactive effects of acetylcholine to a similar level in normotensive and hypertensive rats suggesting that in this form of hypertension, an increased EDHF production, but not a decreased NO, occurs [30].

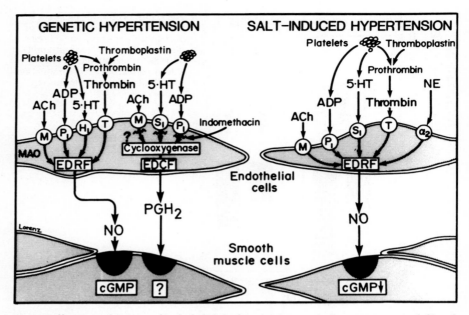

Fig. 4. Different mechanisms of endothelial dysfunction in genetic (i.e., spontaneous; *left*) and salt-induced hypertension of the rat (*right*). *ACh,* acetylcholine; *ADP,* adenosine diphosphate; *cGMP,* cyclic 3′,5′-guanosine monophosphate; *NE,* noradrenaline; *5-HT,* 5-hydroxytryptamine (serotonin); *PGH₂,* prostaglandin H$_2$; EDCF, endothelium-derived contracting factor. (Modified from [27])

Further studies carried out in our lab have demonstrated that the accumulation of cyclic GMP in mesenteric resistance arteries is similar in WKY and SHR [24]. Moreover, the NO-dependent vasodilator tone, assessed by the blood pressure effects of L-NAME, is not higher in hypertensive rats as would be expected in a situation in which the production of NO is increased. The capacity of vascular smooth muscle cells of hypertensive rats to respond to NO, on the other hand, must be fully maintained as organic nitrates lower blood pressure in a similar fashion in both strains of rats and relaxations to sodium nitroprusside are enhanced in this condition [31]. All together, these findings suggest that endogenously-produced NO, otherwise increased in SHR, is unable to properly raise cyclic GMP levels in the vascular smooth muscle cells of these animals. Hence, it appears that an additional event takes place that prevents NO to accomplish its normal hemodynamic functions. The hypertrophied and fibrotic intimal layer of hypertensive vessels [32] may represent a physical barrier for NO accounting for the blunted hemodynamic actions of NO. Also the chemical environment that NO encounters, such as oxidative stress can determine its fate as the results from Nakazono et al. [33] and Tschudi et al. suggest [34].

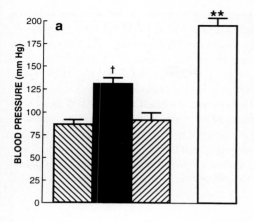

Fig. 5a,b. Blood pressure (**a**) and activity of constitutive NO synthase (cNOS; **b**) in hearts of spontaneously hypertensive *(SHR)* and normotensive Wistar-Kyoto rats *(WKY)* at age 3 weeks *(young)* and 18 weeks *(adult)*. The increased activity of cNOS appears to be related to the hypertensive status of of the animals and not to age or strain differences. *,** $p < 0.05$ and $p < 0.01$ versus normotensives, $p < 0.01$ versus young animals. (From [28])

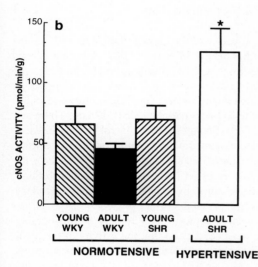

Nitric Oxide in the Hypertensive Heart

The heart is an important target in hypertension which causes hypertrophy, myocardial damage and left ventricular failure [35]. Recent studies have shown that in spontaneous hypertension, the production of NO in this organ is increased. The release of NO from isolated coronary vessels is augmented in the SHR [36]. Moreover, adult SHR possess a higher activity of eNOS in the heart endothelial cells than their normotensive counterparts. Very young prehypertensive SHRs have, in contrast, lower eNOS activity than the normotensive, indicating that the increased activity of eNOS in these cells is indeed related to hypertension (Fig. 5) [28]. Later studies showed that the augmented activity of eNOS in the hypertensive heart happens at the expense of the left ventricle where the highest differential pressure in the cardiovascular system is found [28]. Finally, the expression of endothelial NOS is higher in hypertensive animals [24]. These observations further support the concept that high blood pressure upregulates eNOS and hence the production of NO. It appears that cardiac eNOS activity remains unchanged within the normotensive blood pressure range, and that there is a pressure threshold above which upregulation of the enzyme takes place.

An enhanced production of NO in the hypertensive heart probably acts as a compensatory mechanism by decreasing myocardial contractility and causing vascular dilatation. Nitric oxide in the hypertensive heart may also protect from hypertrophy. High blood pressure causes cardiac hypertrophy and fibrosis which often leads to left ventricular failure [35]. Nitric oxide, which is also a potent inhibitor of smooth muscle cell growth and migration [17], might protect the heart from these deleterious effects of hypertension.

Nitric Oxide and the Kidney in Hypertension

A large number of studies evidence the capacity of the kidney to produce NO and its relevant role in renal function [37–39]. It has been demonstrated that the kidney is very sensitive to the reduction of NO as low doses of NOS inhibitors reduce sodium and water excretion without affecting renal hemodynamics or systemic arterial pressure [37]. Within the kidney, the medulla, which is importantly involved in pressure-induced natriuresis [38], produces most of the NO-dependent cyclic GMP [38]. There are increasing grounds for believing that NO participates in the control of renal medullary blood flow. Firstly, NOS inhibitors decrease medullary, but not cortical blood flow, when selectively applied in the medullary interstitial space and this is associated with sodium retention and development of hypertension [38]. Also, treatment with these inhibitors blunts the pressure-natriuresis response by impairing the renal medullary vasodilation produced by increasing arterial pressure [39].

We have recently presented evidence showing that eNOS activity is considerably higher in the renal medulla than in the renal cortex and that the activity in the renal medulla is also much higher than in the heart and aorta (Fig. 6) [29]. These results indicate that the renal medulla has a greater potential to produce NO compared to the renal cortex or cardiovascular tissues involved in blood pressure regulation. Because NO appears to be highly relevant for renal function one might think that minimal alterations in the renal NO production, which do not affect systemic vascular tone, lead to hypertension because of an impaired regulation of sodium and water excretion [37]. Studies performed in the SHR show that the activity of NOS in the medulla is higher than in the normotensive counterparts, thus further supporting the idea that NO production is enhanced to compensate high arterial pressure (Fig. 6) [29].

In renovascular hypertension, endothelium-dependent vascular relaxation is impaired suggesting a diminished production of EDRF [40]. However, Sigmon and Beierwaltes have demonstrated in a model of renovascular hypertension that treatment with an inhibitor of NO synthesis affects similarly clipped and non clipped kidneys as well as normotensive controls [41]. They concluded that in this model of hypertension, the

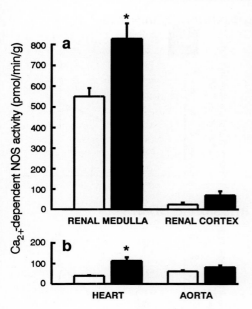

Fig. 6a,b. Ca^{2+}-dependent NO synthase *(NOS)* activity in the kidney (**a**) and the heart and aorta (**b**) in normotensive Wistar-Kyoto rats (WKY, *white bars*) and spontaneously hypertensive rats (SHR, *black bars*). In both strains of rats the activity was much higher in the renal medulla. In the medulla and heart of the SHR the activity was significantly higher than in the same tissues of the WKY. *$p < 0.05$ compared to the same kind of tissue of the WKY. (From [29])

endothelium is not disfunctional but is a critical component in the adaptation to increased blood pressure. This has been corroborated by recent evidence from Dubey et al. who measure higher amounts of cyclic GMP and NO metabolites in serum in kidney-one clip hypertensive rats compared to sham operated [42].

Nitric Oxide in Salt-Sensitive Hypertension

It was reported 10 years ago that Dahl salt-sensitive rats have impaired endothelium-dependent relaxations (Fig. 4) [43]; however, no release of vasoconstrictor prostanoids could be demonstrated as in the SHR [43]. This suggested that a decreased EDRF production contributed to the pathogenesis of this form of hypertension. Chen and Sanders show interesting data supporting this idea [44, 45]. Inhibition of NO synthesis causes a higher increase in blood pressure in normotensive rats (Sprague-Dawley and Dahl salt-resistant) than in Dahl salt-sensitive rats, suggesting that the synthesis of NO is lower in these animals [44, 45]. They also demonstrate that NO production, as assessed by the pharmacological effects of NO inhibitors, improves by increasingly feeding NaCl to salt-resistant Dahl rats [44]. Moreover, administration of L-arginine is effective in lowering blood pressure in the salt-sensitive strain, but not the salt-resistant or the SHR (Fig. 7) [44]. In this line Ikeda et al. demonstrated that eNOS activity is lower in kidneys from salt-sensitive Dahl rats as compared to salt-resistant [46]. Hayakawa et al. have performed interesting experiments demonstrating that renal blood vessels from Dahl salt-sensitive rats display a reduced response to acetylcholine which is accompanied by a decreased release of NO [30]. Studies on the biology of NO in

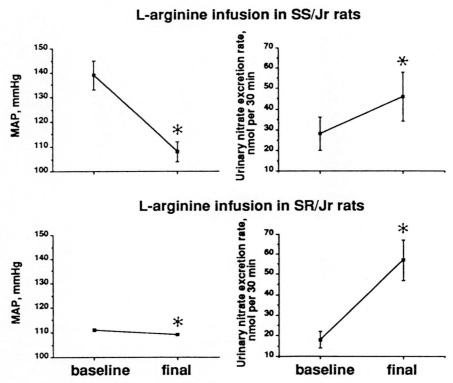

Fig. 7. Effects of L-arginine on mean arterial pressure and urinary nitrate in salt-sensitive and sal-resistant Dahl rats. (Modified from [45])

other models of salt-sensitive hypertension have provided results which point to similar directions. Blood vessels from DOCA salt-sensitive hypertensive rats elicit impaired endothelium-dependent relaxations [47] and cGMP accumulation is diminished [48] suggesting that a decreased production of NO may be involved. Hayakawa and coworkers have demonstrated that these impaired relaxations are paralleled by a diminished release of NO from perfused kidney vessels NO [30]. Beneficial effects of treatment with L-arginine on endothelial function have also been shown [49,50] in this model of hypertension. Rees et al. have recently reported in another model of salt-sensitive hypertension, the Sabra hypertensive-prone rat, that the activity of cNOS from aortic endothelial cells and the concentration of nitrate in plasma is lower than in the Sabra hypertensive resistant rat (see chapter: "Nitric Oxide and Hypertension: Physiology and Pathophysiology" by P. Vallance and S. Moncada) [51]. Studies in humans reveal that the ability of L-arginine to produce NO is diminished in salt-sensitive patients when being salt-loaded [52], thus debating the possibility of treating salt-sensitive hypertension with L-arginine as the studies in animal models suggest [44, 45].

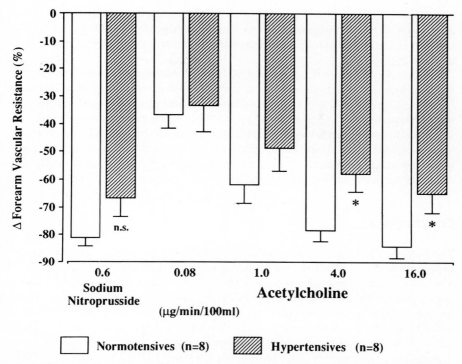

Fig. 8. Changes in forearm vascular resistance caused by acetylcholine and sodium nitroprusside in normotensives and hypertensives. The response to acetylcholine, but not that to nitroprusside was impaired in hypertensive patients. $^*p < 0.05$. (From [54])

Nitric Oxide in Human Hypertension

Experiments in humans, have demonstrated a diminished basal and stimulated NO production (Fig. 8) [53, 54]. The decrease in forearm blood flow induced by L-NMMA, is smaller in hypertensive than in normotensive patients [53]. Most studies, but not all, find endothelium-dependent vasodilations reduced in patients with primary or secondary hypertension [55–57]. The impaired endothelial response in hypertensives can be enhanced by indomethacin, suggesting that vasoconstrictor prostanoids also contribute to impaired endothelium-dependent relaxation in hypertensive patients [56]. Recent studies, where analytical measurements of NO have been performed, suggest that the ability to produce NO may be indeed deterred in human hypertensives [58].

Pulmonary Hypertension

Nitric oxide is present in exhaled air of animals [59] and humans [60]. It is an inhibitory modulator of the pulmonary hypoxic pressor response [61]. Furthermore, NO plays a role in the pulmonary circulation and in the normal oxygenation of the blood through regulation of the ventilation-perfusion matching [61]. It is likely that the NO involved in the physiological regulation of the pulmonary circulation is synthesized by eNOS [59].

On the basis of these findings, inhaled NO has been used for the treatment of various forms of pulmonary hypertension such as the pulmonary hypertension of the newborn [62,63], pulmonary hypertension of congenital heart disease [64], idiopathic pulmonary hypertension [65], acute pneumonia [66] and severe airways disease [67]. It is important to realize that inhaled NO is still at the experimental stage, as the potential hazards of this treatment have not been fully explored. If further data confirm the early promise of this novel treatment, it will not be long before NO delivery and monitoring machines are in routine use on neonatal and intensive care units.

Conclusions

The biology of NO varies widely in different forms of hypertension. In spontaneous hypertension an increased NO production has been demonstrated in the heart, kidney and resistance vessels. However, NO appears to be inefficacious to relax smooth muscle cells probably due to increased oxidative degradation. In renovascular hypertension, NO production also seems to be enhanced. However, in salt-sensitive hypertension and in human hypertension, NO synthesis appears to be diminished. It is likely that in those forms of hypertension where an increased production of NO has been demonstrated, the cause of hypertension is not related to the NO pathway and NO release is increased as a compensatory mechanism to lower blood pressure. On the other hand, in other forms of hypertension, a diminished ability of the endothelium to produce NO might be involved in the pathogenesis of this disease.

References

1. Moncada S, Gryglewski R, Bunting S, Vane JR (1976) An enzyme isolated from arteries transforms prostaglandin endoperoxides to an unstable substance that inhibits platelet aggregation. Nature 263: 663–665
2. Furchgott RF, Zawadzki JV (1980) The obligatory role of endothelial cells in the relaxation of arterial smooth muscle by acetylcholine. Nature 28: 373–376
3. Palmer RMJ, Ferrige AG, Moncada S (1987) Nitric oxide release accounts for the biological activity of endothelium-derived relaxing factor. Nature 327: 524–526
4. Mombouli J-V, Illiano S, Nagao T, Scott-Burden T, Vanhoutte PM (1992) Potentiation of endothelium-dependent relaxations to bradykinin by angiotensin I converting enzyme inhibitors in canine coronary artery involves both endothelium-derived relaxing and hyperpolarizing factors. Circ Res 71: 137–144
5. Lüscher TF, Tanner FC (1993) Endothelial regulation of vascular tone and growth. Am J Hypertens 6: 283S-293S
6. Moncada S, Higgs A (1993) Mechanisms of disease. The L-arginine-nitric oxide pathway. N Engl J Med 329(27): 2002–2012
7. Moncada S (1992) The L-arginine: nitric oxide pathway. Acta Physiol Scand 145: 201–227
8. Sessa WC, Pritchard K, Seyedi N, Wang J, Hintze TH (1994) Chronic exercise in dogs increases coronary vascular nitric oxide production and endothelial cell nitric oxide synthase gene expression. Circ Res 74: 349–353

9. Schmidt HHHW, Zernikow B, Baeblich S, Böhme E (1990) Basal and stimulated formation and release of L-arginine-derived nitrogen oxides from cultured endothelial cells. J Pharmacol Exp Ther 254: 591–597

10. Weiner CP, Lizasoain I, Baylis SA, Knowles RG, Charles IC, Moncada S (1994) Induction of calcium-dependent nitric oxide synthase by sex hormones. Proc Natl Acad Sci USA 91: 5212–5216

11. Rees DD, Palmer RMJ, Moncada S (1989) Role of endothelium-derived nitric oxide in the regulation of blood pressure. Proc Natl Acad Sci USA 86: 3375–3378

12. Vallance P, Collier J, Moncada S (1989) Effects of endothelium derived nitric oxide on peripheral arteriolar tone in man. Lancet ii: 997–1000

13. Radomski MW, Palmer RMJ, Moncada S (1990) An L-arginine to nitric oxide pathway in human platelets regulates aggregation. Proc Natl Acad Sci USA 87: 5193–5197

14. Takase H, Moreau P, Küng CF, Nava E, Lüscher TF (1996) Antihypertensive therapy improves endothelium-dependent relaxation of resistance arteries in nitric oxide deficient hypertension: Effect of verapamil and trandolapril. Hypertension 27: 25–31

15. Palmer RMJ, Bridge L, Foxwell NA Moncada S (1992) The role of nitric oxide in endothelial damage and its inhibition by glucocorticoids. Br. J. Pharmacol 105: 11–12

16. Blot S, Arnal JF, Xu Y, Gray F, Michel J-B (1994) Spinal cord infarcts during long-term inhibition of nitric oxide synthase in rats. Stroke 25: 1666–1673

17. Dubey RK, Lüscher TF (1993) Nitric oxide inhibits angiotensin II-induced migration of vascular smooth muscle cells Hypertension 22:412

18. Lüscher TF, Vanhoutte PM (1986) Endothelium-dependent contractions to acetylcholine in the aorta of the spontaneously hypertensive rat. Hypertension 8: 344–348

19. Dohi Y, Thiel MA, Buhler FR, Luscher TF (1990) Activation of the endothelial L-arginine pathway in pressurized mesenteric resistance arteries: effect of age and hypertension. Hypertension 15: 170–179

20. Lüscher TF (1991) Endothelium-derived nitric oxide: the endogenous nitrovasodilator in the human cardiovascular system. Eur Heart J 12 (Suppl E): 2–11

21. Tschudi MR, Criscione L, Lüscher TF (1991) Effect of aging and hypertension on endothelial function of rat coronary arteries. J Hypertension 9: 164–165

22. Nava E, Wiklund NP, Salazar FJ (1996) Changes in nitric oxide release in vivo in response to vasoactive substances. Br J Pharmacol 119: 1211–1216

23. Nava E, Leone AM, Wiklund NP, Moncada S (1994) Detection of release of nitric oxide by vasoactive substances in the anaesthesized rat. In: Feelisch M, Busse R, Moncada S (eds) The biology of nitric oxide. Portland Press, London, pp 179–181

24. Nava E, López-Farré A, Casado S, Moreau P, Cosentino F, Lüscher TF, The biology of nitric oxide in spontaneous hypertension (in revision)

25. Buga GM, Gold ME, Fukuto, JM, Ignarro L (1991) Shear stress-induced release of nitric oxide from endothelial cells grown on beads. Hypertension. 17: 187–193

26. Sessa WC, Pritchard K, Seyedi N, Wang J, Hintze TH (1994) Chronic exercise in dogs increses coronary vascular nitric oxide production and endothelial cell nitric oxide synthase gene expression. Circ Res 74: 349–353

27. Lüscher TF, Vanhoutte PM (1988) Mechanisms of altered endothelium-dependent responses in hypertensive blood vessels. In: Vanhoutte PM (ed) Relaxing and contracting factors. Humana Press, Clifton, New Jersey, pp 495–509

28. Nava E, Noll G, Lüscher TF (1995) Increased activity of constitutive nitric oxide synthase in cardiac endothelium in spontaneous hypertension. Circulation 91: 2310–2313

29. Nava E, Llinás MT, González JD, Salazar FJ (1996) Nitric oxide synthase activity in renal cortex and medulla of normotensive and spontaneously hypertensive rats. Am J Hypertension 9: 1236–1239

30. Hayakawa H, Hirata Y, Suzuki E, Sugimoto T. Matsuoka H, Kikuchi K, Nagano T. Hirobe M, Sugimoto T (1993) Mechanisms for altered endothelium dependent vasorelaxation in isolated kidneys from experimental hypertensive rats. Am J Physiol 264: H1535 H1541

31. Diederich D, Yang Z, Bühler FR, Lüscher TF (1990) Impaired endothelium-dependent relaxations in hypertensive resistance arteries involve cyclooxigenase pathway. Am J Physiol 258: H445-H451

32. Lindop GBM (1994) Textbook of hypertension. In: Swales JD (ed) Blackwell Scientific, London, pp 663–669

33. Nakazono K, Watanabe N, Matsuno K, Sasaki J, Sato T (1991) Does superoxide underlie the pathogenesis of hypertension? Proc Natl Acad Sci USA 88: 10045–10048
34. Tschudi MR, Mesaros S, Lüscher TF, Malinski T (1996) Direct in situ measurement of nitric oxide in mesenteric resistance vessels. Increased decomposition by superoxide in hypertension. Hypertension 27: 32–35
35. Anversa P, Peng Li, Malhotra A, Zhang X, Herman MV, Capasso J (1993) Effects of hypertension and coronary constriction on cardiac function, morphology, and contractile proteins in rats. Am J Physiol 265: 713–725
36. Kelm M, Feelisch M, Deussen A, Strauer BE, Scharader J (1991) Release of endothelium derived nitric oxide in relation to pressure and flow. Cardiovasc Res 25: 186–193
37. Salazar FJ, Alberola A, Pinilla JM, Romero JC, Quesada T (1993) Salt-induced increase in arterial pressure during nitric oxide synthesis inhibition. 22: 49–55
38. Cowley AW, Mattson DL, Lu S, Roman RJ (1995) The renal medulla and hypertension. Hypertension 25: 663–663
39. Fenoy FJ, Ferrer P, Carbonell L, Salom MG (1995) Role of nitric oxide on papillary blood flow and pressure natriuresis. Hypertension 25: 408–414
40. Lockette W, Otsuka Y, Carretero OA (1986) The loss of endothelium-dependent vascular relaxation in hypertension. Hypertension 8 (Suppl II): II-61-II-66
41. Sigmon DH, Beierwaltes WH (1994) Nitric oxide influences blood flow distribution in renovascular hypertension. Hypertension 23: I-34-I-39
42. Dubey RK, Boegehold MA, Gillespie DG and Rosselli M (1996) Increased nitric oxide activity in early renovascular hypertension. Am J Physiol 270: R118-R124
43. Lüscher TF, Raij L, Vanhoutte PM (1987) Endothelium-dependent vascular responses in normotensive and hypertensive Dahl rats. Hypertension 9: 157–163
44. Chen PY, Sanders PW (1991) L Arginine abrogates salt sensitive hypertension in Dahl/Rapp rats. J Clin Invest 88: 1559 1567
45. Chen PY, Sanders PW (1993) Role of nitric oxide synthesis in salt sensitive hypertension in Dahl/Rapp rats. Hypertension 22: 812 818
46. Ikeda Y, Saito K, Kim JI, Yokoyama M (1995) Nitric oxide synthase isoform activities in kidney of Dahl salt-sensitive rats. Hypertension 26: 1030–1034
47. Voorde JV, Leusen I (1986) Endothelium dependent and independent relaxation of aortic rings from hypertensive rats. Am J Physiol 250: H711 H717
48. Otsuka Y, Dipiero A, Hirt E, Brennaman B, Lockette W (1988) Vascular relaxation and cGMP in hypertension. Am J Physiol 254: H163 69
49. Hayakawa H, Hirata Y, Suzuki E, Kimura K, Kikuchi K, Nagano T, Hirobe M, Omata M (1994) Long term administration of L-Arginine improves nitric oxide release from kikney in deoxycorticosterone acetate salt hypertensive rats. Hypertension 23: 752 756
50. Laurant P, Demolombe B, Berthelot A (1995) Dietary L-Arginine attenuates blood pressure in mineralocorticoid salt hypertensive rats. Clin Exper Hypertension 17: 1009 1024
51. Rees DD, Ben-Ishay D, Moncada S (1996) Nitric oxide and the regulation of blood pressure in the Hypertension-Prone and Hypertension-Resistant Sabra rat. Hypertension 28: 367–371
52. Higashi Y, Oshima T, Watanabe M, Matsuura H, Kajiyama G (1996) Renal response to L-arginine in salt-sensitive patients with essential hypertension. Hypertension 27: 643–648
53. Calver A, Collier J, Moncada S, Vallance P (1992) Effect of local intra-arterial NG-monomethyl-L-arginine in patients with hypertension: the nitric oxide dilator mechanism appears abnormal. J Hypertens 10: 1025–1031
54. Linder L, Kiowski W, Bühler FR, Lüscher TF (1990) Indirect evidence for the release of endothelium-derived relaxing factor in human forearm circulation in vivo: blunted response in essential hypertension. Circulation 81: 1762–1767
55. Panza JA, Casino PR, Kilcoyne CM, Quyyumi AA (1993) Role of endothelium-dependent vascular relaxation of patients with essential hypertension. Circulation 87: 1468–1474
56. Taddei S, Virdis A, Mattei P, Salvetti A (1993) Vasodilation to acetylcholine in primary and secondary forms of human hypertension. Hypertension 21: 929–933
57. Cockcroft JR, Chowienczyk PJ, Benjamin N, Ritter JM. Preserved endothelium-dependent vasodilation in patients with essential hypertension. N Engl J Med (1994) 330: 1036–1040
58. Benjamin N, Copland M, Smith LM (1995) Reduced nitric oxide synthesis in essential hypertension. Endothelium 3: s94
59. Gustafsson SE, Leone AM, Persson MG, Wiklund NP, Moncada S (1991) Endogenous nitric oxide is present in the exhaled air of rabbits, guinea pigs and humans. Biochem Biophys Res Commun 181: 852–857

60. Leone AM, Gustafsson SE, Francis PL, Persson MG, Wiklund NP, Moncada S (1994) Nitric oxide is present in exhaled breath in humans: direct GC-MS confirmation. Biochem Biophys Res Commun 201: 883–887
61. Wiklund NP, Persson MG, Gustafsson LE, Moncada S, Hedqvist P (1990) . Eur J Pharmacol 185: 123–124
62. Roberts JD, Polanger DM, Lang P, Zapol WM (1992) Inhaled nitric oxide in persistentpulmonary hypertension of the newborn. Lancet 340: 818–819
63. Kinsella JP, Neish SR, Shaffer E, Arman SH (1992) Low dose inhalational nitric oxide in persistent pulmonary hypertension of the newborn. Lancet 340: 819–820
64. Roberts JD, Lang P, Bigatello LM, Vlahakes GJ, Zapol WM (1993) Inhaled nitric oxide in congenital heart disease. Circulation 87: 447–453
65. Pepke-Zaba J, Higenbottam TX, Tuan Dinh-Xuan A, Stone D, Wallwork J (1991) Inhaled nitric oxide as a cause of selective pulmonary vasodilatation in pulmonary hypertension. Lancet 338: 1173–1174
66. Blomqvist H, Wickerts CJ, Andreen M, Ullberg U, Ortqvist A, Frostell C (1993) Enhanced pneumonia resolution by inhalation of nitric oxide. Acta Anaesthesiol Scand 37: 110–114
67. Adatia I, Thomson J, Landzberg M, Wessel D (1993) Inhaled nitric oxide in chronic obstructive lung disease. Lancet 341: 307–308

Endothelial Alterations in Atherosclerosis: The Role of Nitric Oxide

J. P. Cooke and P. S. Tsao

Summary

Hypercholesterolaemia and other disorders that predispose to atherogenesis are all associated with reduced NO activity. The reduction of NO activity may be due to reduced NO synthesis and/or increased degradation. A reduction in NO activity favors vasoconstriction, platelet adherence and aggregation, monocyte adherence, and generation of superoxide anion. NO affects these processes by cyclic GMP-dependent and cyclic GMP-independent pathways. By regulating oxidant-responsive transcriptional mechanisms, nitric oxide inhibits the expression of molecules involved in atherogenesis such as vascular cell adhesion molecule-1 and monocyte chemotactic protein-1.

Enhancement of endogenous NO activity may be a new therapeutic strategy for preventing the progression of atherosclerotic vascular disease, restenosis, and thrombosis.

Endothelial Alterations in Atherogenesis

Within 1 week of exposure to a high fat diet, monocytes can be seen clinging to the surface of the endothelium of the aorta in hypercholesterolaemic animals [1]. This monocyte-endothelial cell interaction is due to the expression by the endothelium of specific adhesion molecules (such as vascular cell adhesion molecule or VCAM-1), as well as chemokines (such as monocyte chemotactic protein or MCP-1) [2–4]. The monocytes typically enter the sub-endothelial space by forcing their way between two adjacent endothelial cells. Once in the sub-endothelial space the monocytes become tissue macrophages, scavenging oxidized LDL cholesterol. The oxidized LDL cholesterol is taken up via "scavenger" receptors expressed on the surface of the monocyte [1]. The expression of these receptors is not down-regulated by intracellular levels of cholesterol, permitting the macrophages to imbibe prodigious amounts of oxidized lipoprotein. In the process, the tissue macrophage becomes a foam cell, laden with lipid. Collections of these foam cells form the first grossly visible lesions of atherosclerosis – the fatty streak.

Fatty streaks tend to occur at bends, branches, and bifurcations in the conduit vessels [5, 6]. In these regions of the vessel there is evidence of vortical flow. The shear stress which the endothelium experiences in these areas is low, or even reversed [7].

In these areas of low shear stress, residence particle time is increased favoring inter-action of lipoproteins, as well as monocytes, with the vessel wall. Perhaps more important, these alterations in local haemodynamic forces exert significant effects upon endothelial biology.

Chronic high levels of laminar shear stress in the physiological range down-regu-late the expression of vascular cell adhesion molecule [8,9] and monocyte chemotactic protein-1 [10]; by contrast, shear stress up-regulates the expression of nitric oxide syn-thase [11] as well as superoxide dismutase [12]. Accordingly, endothelial cells exposed to physiological levels of shear stress are capable of generating relatively more nitric oxide, and scavenging more superoxide anion, than endothelial cells in low shear stress regions. Furthermore, shear stress triggers the release of nitric oxide and prostacyclin, both of which inhibit monocyte adherence [11, 13, 14]. These effects of shear stress on endothelial biology explain the observation that areas of the vessel exposed to high laminar flow tend to be disease-free; by contrast, atherosclerotic lesions tend to develop at bends or bifurcations where flow is disturbed.

Nitric Oxide – A Pluripotential Factor

By 1988, it had become clear that endothelium-derived nitric oxide was not only an important vasodilator, but could inhibit many other cellular processes. A number of *in vitro* studies had demonstrated that endogenous or exogenous nitric oxide could inhibit leukocyte adherence to the endothelium [15, 16], platelet adherence [19] and aggregation [17, 18], as well as vascular smooth muscle proliferation [19]. Because these processes are all involved in atherogenesis, we began to formulate the concept that nitric oxide may be an endogenous anti-atherogenic molecule. We speculated that under normal conditions, nitric oxide would exert a brake upon these processes. How-ever, under conditions where nitric oxide activity was reduced, atherogenesis would be favored.

Indeed, a number of investigators had previously demonstrated that nitric oxide activity was reduced in hypercholesterolaemic animals and man [20–23]. In organ chamber studies, Verbeuren and colleagues illustrated that endothelium-dependent relaxation was attenuated in vascular rings of the thoracic aortae in hypercholesterol-aemic animals [20]. Ludmer and co-workers extended these results to humans [23]; in their studies of coronary vascular reactivity in vivo, acetylcholine stimulated endo-thelium-dependent relaxation in epicardial coronary arteries but induced a pronoun-ced vasoconstriction in the coronary vessel of patients with atherosclerosis.

In 1988, at the Copper Mountain Symposium, Moncada announced his discovery that L-arginine was the precursor for endothelium-derived nitric oxide. This observa-tion stimulated us to test our hypothesis that nitric oxide was an endogenous anti-atherogenic molecule. We postulated that if nitric oxide activity could be enhanced in a sustained fashion, atherogenesis might be inhibited.

To determine if nitric oxide activity could be enhanced in hypercholesterolaemic animals, the following experiment was performed [25]. New Zealand white rabbits were fed a normal or high cholesterol diet for 10 weeks. Subsequently, animals were anaesthetised for studies of hind limb blood flow. Intra-arterial infusions of acetylcholine induced dose-dependent vasodilatation which was attenuated in the hypercholesterolaemic animals. In contrast, hind limb response to sodium nitroprusside was similar in the normal and hypercholesterolaemic animals.

Subsequently, the animals received an intravenous infusion of saline, L-arginine, or D-arginine. After the intravenous infusion, the hind limb blood flow studies were repeated. The infusion of L-arginine augmented hind limb blood flow responses to acetylcholine in the hypercholesterolaemic animals but did not affect the response in normal animals. The L-arginine infusion did not affect responses to sodium nitroprusside. Infusion of saline alone did not affect vascular responses. D-Arginine infusions did not mimic the effect of L-arginine.

Subsequently, the thoracic aortae was removed from these animals for *in vitro* studies [26]. Thoracic aortae obtained from hypercholesterolaemic animals receiving the saline infusion manifested an impairment of endothelium-dependent vasorelaxation in response to acetylcholine. By contrast, those animals that received L-arginine exhibited a normalization of endothelium-dependent vasorelaxation. D-arginine did not mimic the affects of L-arginine. These studies led us to conclude that the endothelial vasodilator dysfunction observed in hypercholesterolemia was reversible by administration of the NO precursor. Subsequent studies revealed that the endothelial vasodilator dysfunction in hypercholesterolaemic humans was also reversible by L-arginine infusions [27–29].

Enhancement of Nitric Oxide Synthesis Inhibits Atherogenesis

We reasoned that if NO activity could be restored chronically, it would inhibit atherogenesis. To test this hypothesis, New Zealand white rabbits were fed normal chow or a high cholesterol diet [30]. Half of the hypercholesterolaemic animals also received L-arginine in their drinking water (2.25 grams%). This supplementation of dietary arginine caused plasma arginine levels to double, but did not affect the lipoprotein profile, nor did it affect intraarterial pressures. At the end of 10 weeks, the thoracic aortae were harvested for studies of vascular reactivity and histomorphometry.

As expected, endothelium-dependent vasorelaxation in response to acetylcholine was inhibited in the vascular rings derived from hypercholesterolaemic animals receiving vehicle alone. By contrast, there was a partial restoration of endothelium-dependent vasodilation in the arginine-treated hypercholesterolaemic rabbits. This partial restoration of endothelial vasodilator function was associated with a striking effect upon vascular structure. In the hypercholesterolaemic animals receiving vehicle, 40% of the thoracic aorta was involved by lesion as assessed by planimetry. By contrast, only 9% of the thoracic aorta was involved by lesions in the hypercholesterolaemic animals receiving arginine.

These data supported the hypothesis that endothelium-derived nitric oxide is an endogenous antiatherogenic molecule. Subsequently, we undertook studies to elucidate the mechanisms by which the arginine–NO system exerted these effects.

Mechanisms by Which Nitric Oxide Inhibits Atherogenesis

We hypothesized that endogenous nitric oxide exerted its anti-atherogenic effects in part by inhibiting endothelial adhesiveness for monocytes. To test this hypothesis, animals were fed a normal or high cholesterol chow for 2 weeks [31]. At this time point, the thoracic aortae of the hypercholesterolaemic animals are grossly normal with no apparent lesions. The thoracic aortae were harvested and segments prepared for measurements of nitric oxide by chemiluminescence. Adjacent rings were prepared for a functional binding assay. Some of the hypercholesterolaemic animals received L-arginine (2.25 grams%) in their drinking water in an effort to enhance NO synthesis. By contrast, some of the normal animals received nitroarginine (13.5 mg/kg), also administered in the drinking water, in an attempt to reduce NO synthesis. We hypothesized that modulation of NO activity would alter endothelial adhesiveness.

Vascular rings were exposed to calcium ionophore and the conditioned medium was collected for measurement of nitrogen oxides. The generation of nitrogen oxides in rings from hypercholesterolaemic animals was not different from that observed in rings from normocholesterolaemic animals. By comparison, there was increased generation of nitrogen oxides by vascular rings obtained from hypercholesterolaemic animals that had received supplemental arginine. There was a significant reduction in nitrogen oxides produced by vascular rings obtained from normocholesterolaemic animals treated with nitroarginine.

These alterations in NO activity had dramatic effects upon endothelial adhesiveness for monocytes in the functional binding assay. In the functional binding assay, vascular rings were dissected and laid flat in a Petri dish containing HBSS medium. Monocytoid cells (WEHI 78/24 cells) were added to the medium, and the preparation was placed on a rocker platform for 15 min. Subsequently, non-adherent cells were washed off the endothelial surface with fresh medium. The tissues were fixed in gluteraldehyde, and the adherent monocytes counted by videomicroscopy. Monocyte adherence was increased three-fold in rings obtained from hypercholesterolaemic animals. The number of monocytes bound to the endothelium was significantly reduced in vascular rings obtained from hypercholesterolaemic animals treated with arginine. By contrast, monocyte binding was dramatically increased to the endothelium of vascular rings obtained from normocholesterolaemic animals treated with nitroarginine. Indeed, there was a 10-fold increase in the numbers of bound monocytes in these tissues; therefore, adhesiveness of these tissues from monocytes was even greater than that observed in tissues obtained from hypercholesterolaemic animals. These studies revealed that modulation of NO activity could have striking effects upon endothelial adhesiveness for monocytes.

Mechanisms by Which Nitric Oxide
Inhibits Endothelial Adhesiveness

In preliminary studies, we have observed an increased expression of monocyte chemotactic protein-1 (MCP-1), in the thoracic aortae obtained from hypercholesterolaemic animals. By Northern analysis, there is no detectable message for MCP-1 in thoracic aortae from normocholesterolemic animals. The administration of L-arginine attenuates the expression of MCP-1 in the thoracic aortae of hypercholesterolaemic animals. Of greater interest, the chronic administration of nitroarginine to normal animals dramatically increases the expression of MCP-1 in the vessel wall. These studies suggest that modulation of NO activity significantly affects the expression of a major chemokine involved in atherogenesis.

Observations from our laboratory and those of others indicates that nitric oxide can exert potent effects upon transcription of molecules regulating endothelial adhesiveness (e.g., MCP-1 and VCAM-1) [9,32,34]. These genes are regulated in part by an oxidant-responsive transcriptional pathway. Nuclear Factor κB (NFκB) is a heterodimer composed of p65 and p50 subunits [36]. Normally, this transcriptional protein exists in the cytoplasm in an inactive state, bound to an inhibitory protein known as IκBα. In a state of oxidative stress, IκBα becomes phosphorylated and disengages from NFκB. It is then degraded, leaving NFκB free to translocate to the nucleus where it activates a number of genes including VCAM-1, MCP-1, and monocyte colony stimulating factor (MCSF) [33, 35, 37]. Oxidized lipoprotein and certain fatty acids are capable of activating this transcriptional pathway. By contrast, nitric oxide appears to inhibit this pathway, in part, by enhancing the activity of the inhibitor IκBα [38].

We speculate that nitric oxide may also inhibit the activation of the transcriptional pathway by directly reducing the endothelial generation of superoxide anion. Nitric oxide may directly inhibit oxidative enzyme activity, possibly by binding to the heme moiety or critical sulfhydryl groups in these enzymes [39, 40]. In this way, nitric oxide may oppose the effect of hypercholesterolaemia, diabetes mellitus and circulating factors such as angiotensin II to enhance endothelial oxidative enzyme activity [41–43].

To further test this hypothesis, we performed the following studies. To enhance the elaboration of endothelium-derived nitric oxide, we exposed cultured endothelial cells to shear stress using a cone plate viscometer. We found that prior exposure to shear stress made endothelial cells resistant to the effects of oxidized lipoprotein or cytokines. In comparison to quiescent cells, shear stress enhanced the production of nitric oxide; this effect was associated with a reduced elaboration of superoxide anion in response to oxidized lipoprotein (oxidized LDL cholesterol, 30 µg/ml), or cytokines (TNFα and lipopolysaccharide). Gel retardation assays revealed that the oxidized lipoprotein or cytokines increased the activity of NFκB in cells previously exposed to quiescent conditions; by contrast, cells previously exposed to shear stress were resistant to cytokine or oxidized lipoprotein-induced NFκB expression.

The expression of VCAM-1 (as assessed by FACS analysis), and the adherence of monocytes to the endothelium, was increased by exposure of the endothelial cells to oxidized lipoprotein or cytokines. The expression of VCAM-1 and the adherence of monocytes to the endothelium were suppressed by prior exposure to shear stress. These effects of shear stress were abrogated by co-incubation with nitroarginine, the antagonist of NO synthase.

These studies support the hypothesis that endothelium-derived nitric oxide inhibits endothelial adhesiveness for monocytes. This effect of nitric oxide appears to be due in part to its suppression of an oxidant-responsive transcriptional pathway mediating the expression of vascular cell adhesion molecule-1 (Fig. 1).

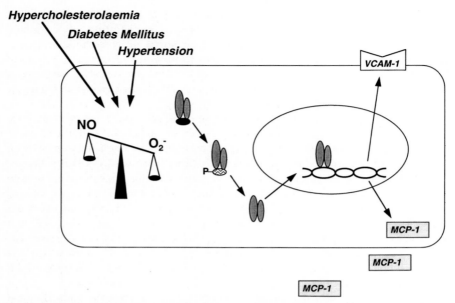

Fig. 1. Hypercholesterolaemia, diabetes mellitus, and hypertension are associated with the generation of oxygen-derived free radicals, and a decreased activity of nitric oxide *(NO)*. This imbalance between superoxide and NO causes an increase in oxidative stress. With an increase in oxidative stress the inhibitory nuclear factor IκBα becomes phosphorylated and degraded. The loss of IκBα allows nuclear factor κB to diffuse into the nucleus and initiate NFκB-mediated transcription of genes such as VCAM-1 and MCP-1

In separate studies, we have confirmed the observation of Zeiher and colleagues that nitric oxide also regulates the expression of MCP-1. Cultured rabbit vascular smooth muscle cells were exposed to lipopolysaccharide or to oxidized LDL cholesterol. Some cells were co-incubated with an NO donor. Subsequently, MCP-1 expression was assessed by Northern analysis. Lipopolysaccharide or oxidized lipoprotein dramatically augmented the expression of MCP-1 by vascular smooth muscle cells; this effect was markedly attenuated by co-incubation with DETA-NO (an NO donor).

These studies indicate that endothelium-derived nitric oxide is capable of interfering with atherogenesis. However, these studies leave open the question as to whether one can inhibit the progression of disease by enhancing NO activity in vessels with pre-existing lesions.

Regression and Progression: Dependency upon Nitric Oxide

To answer this question, the following studies were performed [44]. Animals were fed normal chow or a high cholesterol diet for 10 weeks. At this time point, dietary L-arginine supplementation was provided to half of the animals on the high cholesterol diet.

In both of the hypercholesterolaemic groups, the high fat diet was continued for another 13 weeks. At regular intervals, thoracic aortae were harvested for studies of vascular reactivity and histomorphometry. In animals fed a high cholesterol diet there was a progressive attenuation of endothelium-dependent relaxation. By contrast, in the animals receiving L-arginine, there was an improvement in endothelial vasodilator function at weeks 14 and 18. However, by week 23, the beneficial effects of L-arginine were lost in a majority of the animals and endothelial-dependent vasodilatation was significantly impaired. However, in a few of these animals, L-arginine induced a persistent restoration of endothelial vasodilator function.

The studies of vascular structure were informative. At week 10, approximately 25% of the thoracic aortae was involved by lesions in the hypercholesterolaemic groups. Administration of L-arginine induced an apparent regression of lesions. At week 18, about 30% of the thoracic aortae was involved by plaque in the hypercholesterolaemic animals; by contrast, only 18% of the thoracic aorta was involved by lesions in the arginine-treated hypercholesterolaemic animals. However, by week 23 in the majority of animals, the salutary effects of arginine were lost. At week 23, the average lesion surface area was not different between the hypercholesterolaemic animals treated with vehicle and those treated with arginine. However, in the three animals in which there was a persistent restoration of endothelial vasodilator function, there remained a dramatic attenuation of lesion surface area. In these animals, lesions involved only 5% of the surface area of the thoracic aorta.

These studies suggest that in the setting of pre-existing lesions, enhancement of nitric oxide may be able to induce regression of lesions. Preliminary studies in our laboratory suggest that the mechanism of this regression is in part due to accelerated apoptosis of foam cells.

Enhancement of Nitric Oxide Activity: A New Therapeutic Strategy

These animal studies beg the question of whether modulation of NO activity may be a new therapeutic strategy. Indeed, preliminary studies in humans are encouraging. We, and others, have demonstrated that administration of arginine to humans with risk factors for atherosclerosis or with pre-existing lesions can enhance endothelium-dependent vasodilatation [27–29]. This improvement of endothelial vasodilator function is associated with an effect upon circulating blood elements. As observed in our animal models, oral arginine administration attenuates platelet aggregation in hypercholesterolaemic humans. As in animal models, this is associated with an increase in platelet cyclic GMP, indicating that nitric oxide exerts its effects through nitric oxide. Whether this effect of arginine is due to an enhancement of platelet and/or endothelium-derived nitric oxide remains an open question.

In addition, we have observed that circulating mononuclear cells exhibit an altered behavior in hypercholesterolaemic humans [46]. When these cells are isolated from hypercholesterolaemic humans they exhibit enhanced adhesiveness for endothelial cells in culture. This increase in monocyte adhesiveness is reversed by oral administration of L-arginine [46].

Whether these effects of L-arginine are directly due to its enhancement of NO synthesis remains an open question. However, the most parsimonious explanation for the observations of our group is that L-arginine is exerting its effects via NO. The paradox of these observations is that arginine should not be rate limiting given that the K_m for NO synthase is in the micromolar range and the intracellular concentrations of L-arginine are in the millimolar range [47]. It is most likely that hypercholesterolemia and other states engendering atherogenesis, induce an alteration in enzyme affinity, or arginine availability.

Indeed, it has recently been discovered that the level of an endogenous NO synthase inhibitor is elevated in certain disease states. Asymmetric dimethylarginine (ADMA) is elevated in uraemic patients [48]. In these patients, ADMA circulates in plasma concentrations that are sufficient to antagonise NO synthase and induce vasoconstriction in isolated vascular rings. Circulating levels of ADMA are also elevated in hypercholesterolaemic rabbits [49] as well as patients [50]. with peripheral arterial disease. It is intriguing to speculate that the enhanced elaboration of this endogenous NOS inhibitor is a risk factor for the development of atherosclerosis. Indeed, enzymes responsible for the synthesis and/or degradation of this circulating factor may be a target for drug development.

Acknowledgments. This work was supported in part by a grant from the National Heart, Lung, and Blood Institute (1RO1HL48638) and was done during the tenure of a Grant-in-Aid Award from the American Heart Association and Sanofi Winthrop. Dr. Cooke is a recipient of the Vascular Academic Award from the National Heart, Lung, and Blood Institute (1KO7HC02660), and is an Established Investigator of the American Heart Association.

References

1. Ross R (1986) The pathogenesis of atherosclerosis. N Engl J Med 314:488–500
2. Cybulsky MI, Gimbrone MA Jr (1991) Endothelial expression of a mononuclear leukocyte adhesion molecule during atherogenesis. Science 251:788–91
3. Berliner JA, Navab M, Fogelman AM, Frank JS, Demer LL, Edwards PA, Watson AD, Lusis AJ (1995) Atherosclerosis: basic mechanisms. Oxidation, inflammation, and genetics. Circulation 91:2488–96
4. Butcher EC, Picker LJ (1996) Lymphocyte homing and homeostasis. Science 272(5258):60–66
5. Rokitansky C (1952) The pathological anatomy of the organs of respiration and circulation. In: A manual of pathological anatomy. Vol 4 trans. from German by GE Day. London, Sydenham Society
6. Virchow R (1860) In: Cellular pathology as based upon physiological and pathological histology. Trans. from German by F Clance. Churchill, London
7. Nerem RM (1992) Vascular fluid mechanics, the arterial wall and atherosclerosis. J Biomech Eng 114:274–282
8. Ohtsuka A, Ando J, Korenaga R, Kamiya A, Toyama-Sorimachi N, Miyasaka M (1993) The effect of flow on the expression of vascular adhesion molecule-1 by cultured mouse endothelial cells. Biochem Biophys Res Comm 193:303–310
9. Tsao PS, Buitrago R, Chan JR, Cooke JP (1997) Fluid flow inhibits endothelial adhesiveness: NO and transcriptional regulation of VCAM-1. Circulation (in press)
10. Shyy YJ, Hsieh HJ, Usami S, Chien S (1994) Fluid shear stress induces a biphasic response of human monocyte chemotactic protein 1 gene expression in vascular endothelium. Proc Nat Acad Sci USA 91:4678–82

11. Uematsu M, Ohara Y, Navas JP, Nishida K, Murphy TJ, Alexander RW, Nerem RM, Harrison DG (1995) Regulation of endothelial cell nitric oxide synthase mRNA expression by shear stress. Am J Physiol 269:C1371–8
12. Inoue N, Ramasamy S, Fukikai T, Nerem RM, Harrison DG (1996) Shear stress modulates expression of Cu/Zn superoxide dismutase in human aortic endothelial cells. Circ Res 79:32–37
13. Frangos JA, Eskin SG, McIntire LV, Ives CL (1985) Flow effects on prostacyclin production by cultured human endothelial cells. Science 227: 1477–9
14. Bath PMW, Hassall DG, Gladwin A-M, Palmer RMJ, Martin JF (1991) Nitric oxide and prostacyclin. Divergence of inhibitory effects on monocyte chemotaxis and adhesion to endothelium in vitro. Arterioscler Thromb 11:254–60
15. Lefer AM (1995) Attenuation of myocardial ischemia-reperfusion injury with nitric oxide replacement therapy. Ann Thor Surg 60:847–51
16. Kubes P, Kanwar S, Niu XF, Gaboury JP (1993) Nitric oxide synthesis inhibition induces leukocyte adhesion via superoxide and mast cells. FASEB J 7:1293–9
17. Radomski MW, Palmer RM, Moncada S (1987) The anti-aggregating properties of vascular endothelium: interactions between prostacyclin and nitric oxide. Br J Pharm 92:639–46
18. Radomski MW, Palmer RM, Moncada S (1987) The role of nitric oxide and cGMP in platelet adhesion to vascular endothelium. Biochem Biophys Res Comm 148:1482–9
19. Garg UC, Hassid A (1989) Nitric oxide-generating vasodilators and 8-bromo-cyclic guanosine monophosphate inhibit mitogenesis and proliferation of cultured rat vascular smooth muscle cells. J Clin Invest 83:1774–77
20. Verbeuren TH, Jordaens FH, Zonnekeyn LL, VanHove GE, Coene MC, Herman AG (1986) Effect of hypercholesterolemia on vascular reactivity in the rabbit. Circ Res 58:552–564
21. Freiman PC, Mitchell GG, Heistad DD, Armstrong ML, Harrison DG (1986) Atherosclerosis impairs endothelium-dependent vascular relaxation to acetylcholine and thrombin in primates. Circ Res 58:783–9
22. Cohen RA, Zitnay KM, Haudenschild CC, Cunningham LD (1988) Loss of selective endothelial cell vasoactive functions in pig coronary arteries caused by hypercholesterolemia. Circ Res 63:903–910
23. Ludmer PL, Selwyn AP, Shook TL, Wayne RR, Mudge GH, Alexander RW, Ganz P (1986) Paradoxical vasoconstriction induced by acetylcholine in atherosclerotic coronary arteries. N Engl J Med 15:1046–51
24. Zeiher AM, Drexler H, Wollschlaeger H, Saurbier B, Just H (1989) Coronary vasomotion in response to sympathetic stimulation in humans: Importance of the functional integrity of the endothelium. J Am Coll Cardiol 14:1181–1190
25. Girerd XJ, Hirsch AT, Cooke JP, Dzau VJ, Creager MA (1990) L-arginine augments endothelium-dependent vasodilation in cholesterol-fed rabbits. Circ Res 67:1301–1308
26. Cooke JP, Andon NA, Girerd XJ, Hirsch AT, Creager MA (1991) Arginine restores cholinergic relaxation of hypercholesterolaemic rabbit thoracic aorta. Circulation 83:1057–62
27. Creager MA, Gallagher SJ, Girerd XJ, Coleman SM, Dzau VJ, Cooke JP (1992) L-arginine improves endothelium-dependent vasodilation in hypercholesterolaemic humans. J Clin Invest 90:1248–5123
28. Drexler H, Zeiher AM, Meinzer K, Just H (1991) Correction of endothelial dysfunction in coronary microcirculation of hypercholesterolaemic patients by L-arginine. Lancet 338(8782–8783):1546–50
29. Clarkson P, Adams MR, Powe AJ, Donald AE, McCredie R, Robinson J, McCarthy SN, Keech A, Celermajer DS, Deanfield JE (1996) Oral L-arginine improves endothelium-dependent dilation in hypercholesterolaemic young adults. J Clin Invest 7:1989–94
30. Cooke JP, Singer AH, Tsao PS, Zera P, Rowan RA, Billingham ME (1992) Anti-atherogenic effects of L-arginine in the hypercholesterolaemic rabbit. J Clin Invest 90:1168–72
31. Tsao P, McEvoy LM, Drexler H, Butcher EC, Cooke JP (1994) Enhanced endothelial adhesiveness in hypercholesterolemia is attenuated by L-arginine. Circulation 89:2176–82
32. Zeiher AM, Fisslthaler B, Schray-Utz B, Busse R (1995) Nitric oxide modulates the expression of monocyte chemoattractant protein 1 in cultured human endothelial cells. Circ Res 76:980–86
33. Marui N, Offerman MK, Swerlick R, Kunsch C, Rosen CA, Ahmad M, Alexander RW, Medford RM (1993) Vascular cell adhesion molecule-1 (VCAM-1) gene transcription and expression are regulated through an antioxidant-sensitive mechanism in human vascular endothelial cells. J Clin Invest 92:1866–74

34. DeCaterina R, Libby P, Peng H-B, Thannickal VJ, Rajavashisth TB, Gimbrone MA Jr, Shin WS, Liao JK (1995) Nitric oxide decreases cytokine-induced endothelial activation. J Clin Invest 1995; 96:60–68
35. Satriano JA, Shuldiner M, Hora K, Xing Y, Shan Z, Schlondorff D (1993) Oxygen radicals as second messengers for expression of the monocyte chemoattractant protein, JE/MCP-1, and the monocyte colony-stimulating factor, CSF-1, in response to tumor necrosis factor-alpha and immunoglobulin G. Evidence for involvement of reduced nicotinamide adenine dinucleotide phosphate (NADPH)-dependent oxidase. J Clin Invest 92:1564–71
36. Collins T, Read MA, Neish AS, Whitley MZ, Thanos D, Maniatis T (1995) Transcriptional regulation of endothelial cell adhesion molecules: NF-kappa B and cytokine-inducible enhancers. FASEBJ 9:899–909
37. Peng HB, Rajavashisth TB, Libby P, Liao JK (1995) Nitric oxide inhibits macrophage-colony stimulating factor gene transcription in vascular endothelial cells. J Biol Chem 270:17050–55
38. Peng HB, Libby P, Liao JK (1995) Induction and stabilization of I kappa B alpha by nitric oxide mediates inhibition of NF-kappa B. J Biol Chem 270:14214–19
39. Clancy RM, Leszczynska P, Piziak J, Abramson SB (1992) Nitric oxide, an endothelial cell relaxation factor, inhibits neutrophil superoxide anion production via a direct action on NADPH oxidase. J Clin Invest 90:1116–1121
40. Niu X-F, Smith CW, Kubes P (1994) Intracellular oxidative stress induced by nitric oxide synthesis inhibition increases endothelial cell adhesion to neutrophils. Circ Res 74:1133–1140
41. Ohara Y, Petersen TE, Harrison DG (1993) Hypercholesterolemia increases endothelial superoxide anion production. J Clin Invest 91:2546–51
42. Tesfamariam B, Cohen RA (1992) Free radicals mediate endothelial cell dysfunction caused by elevated glucose. Am J Physiol 263(2 Pt 2):H321-6, 1
43. Griendling KK, Minieri CA, Ollerenshaw JD, Alexander RW (1994) Angiotensin II stimulates NADH and NADPH oxidase activity in cultured vascular smooth muscle cells. Circ Res 74:1141–8
44. Candipan RC, Wang BY, Buitrago R, Tsao PS, Cooke JP (1996) Regression or progression. Dependency on vascular nitric oxide. Arterioscler Thromb Vasc Biol 16:44–50
45. Wolf A, Zalpour C, Theilmeier G, Wang B-Y, Ma A, Anderson B, Tsao PS, Cooke JP. Dietary L-arginine supplementation normalizes platelet aggregation in hypercholesterolaemic humans. Arterioscler Thromb (under review)
46. Theilmeier G, Zalpour C, Ma A, Anderson B, Wang B-Y, Wolf A, Candipan RC, Tsao PS, Cooke JP. Adhesiveness of mononuclear cells in hypercholesterolaemic humans is normalized by dietary arginine. Arterioscler Thromb Vasc Biol (under review)
47. Forstermann U, Schmidt HHHW, Pollock JS, Sheng H, Mitchell JA, Warner TD, Nakane M, Murad F (1991) Isoforms of nitric oxide synthase: Characterization and purification from different cell types. Biochem Pharmacol 42:1849–1857
48. Arese M, Strasly M, Ruva C, Costamagna C, Ghigo D, MacAllister R, Verzetti G, Tetta C, Bosia A, Bussolino F (1995) Regulation of nitric oxide synthesis in uraemia. Neph Dialysis Transplant 10:1386–97
49. Bode-Böger SM, Böger RH, Kienke S, Junker W, Frolich JC (1996) Elevated L-arginine/dimethylarginine ratio contributes to enhanced systemic NO production by dietary L-arginine in hypercholesterolaemic rabbits. Biochem Biophys Res Comm 219:598–603
50. Bode-Böger SM, Böger RH, Alfke H, Heinzel D, Tsikas D, Creutzig A, Alexander K, Frolich JC (1996) L-arginine induces nitric oxide-dependent vasodilation in patients with critical limb ischemia. A randomized, controlled study. Circulation 93:85–90

Other Factors in Endothelial Cell Dysfunction in Hypertension and Diabetes

R. A. COHEN and P. J. PAGANO

Introduction

It is difficult, if not impossible, to explain completely the regulation of endothelial cell vasomotor function, particularly when it is altered by disease processes, simply by alterations in the quantity of nitric oxide released from endothelial cells (Cohen 1995). It is now clear that superoxide anion inactivates nitric oxide, that superoxide anion levels are a major determinant of nitric oxide activity and endothelial cell function, and that destruction of nitric oxide by superoxide anion plays a role in decreased endothelium-dependent relaxation in a number of vascular diseases. In addition, in some forms of vascular disease, vasoconstrictor prostanoids apparently can antagonize the vasodilation that is mediated by nitric oxide, and this may occur by interaction with superoxide. This brief review summarizes current understanding of these "other factors" which contribute to abnormal endothelial cell control of vascular tone in vascular disease states.

Generation and Disposition of Oxygen-Derived Free Radicals in Vascular Tissues

Superoxide anion is normally produced by all tissues as a result of mitochondrial oxidative metabolism. Of importance is the fact that various other enzymes may give rise to superoxide anion as a result of oxidation of NADH or NADPH. Other oxidases, including xanthine oxidase, give rise to superoxide anion. Some of these enzymes have defined roles, such as cytochrome P450 reductase or cyclooxygenase, which give rise to oxidized lipid peroxides and superoxide anion. Others have as yet undefined functions, but have been detected as a result of their ability to generate superoxide anion.

Superoxide is primarily disposed of enzymatically by superoxide dismutase (SOD) (Fridovich 1978). The product of the dismutation reaction is hydrogen peroxide which is decomposed by catalase and glutathione peroxidase. The enzymatic activity of SOD is catalyzed by a family of three proteins in eukaryotes. Cytosolic SOD has a catalytic center that includes a copper and zinc atom. A second form of SOD is a secreted form of the copper/zinc enzyme and is tethered to heparan on the extracellular surface of all cells (Abrahamsson et al. 1992; Carlsson et al. 1995; Stratlin et al. 1995). The third SOD is manganese SOD which is known to be induced by cytokines in a variety of cells and is associated with mitochondria (Marklund 1992).

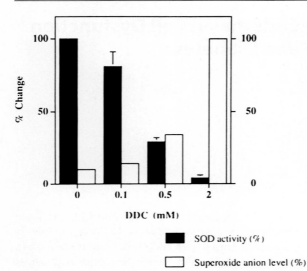

Fig. 1. The effect of increasing concentrations of DDC (0–2 mM) is shown on SOD activity as measured by cytochrome C assay and superoxide anion levels measured by lucigenin chemiluminescence (Pagano et al. 1995; Ito et al. 1996). Data are presented as the percent change in SOD activity *(filled bars)* and superoxide anion levels *(open bars)* measured with no DDC and show the inverse relationship between the two variables

In normal rabbit aorta the majority of SOD activity is the copper/zinc enzyme (Pagano et al. 1995; Stratlin et al. 1995). Under basal conditions, the importance of the activity of this enzyme in the rabbit aorta is made apparent by measuring superoxide anion levels after inhibiting SOD with the copper chelator, diethyldithiocarbamate (DDC) (Pagano et al. 1993). In the experiments shown in Fig. 1, superoxide anion was measured by quantitating the light produced when the chemiluminescent indicator, lucigenin, reacts with superoxide (Faulkner and Fridovich 1993). DDC caused a concentration-dependent rise in the superoxide anion levels which correlated well with inhibition of superoxide dismutase activity in the tissue (Fig. 1). Since under basal conditions, it is difficult to detect any superoxide anion, these studies indicate that copper/zinc SOD scavenges nearly all superoxide being generated in the rabbit aorta.

In the rabbit aorta, rotenone failed to decrease superoxide levels indicating that the superoxide anion generated under basal conditions was not the result of mitochondrial NADH dehydrogenase (Pagano et al. 1995). Rather, the activity could be inhibited by iodonium compounds which bind and inactivate flavin containing enzymes which oxidize NADH and NADPH (Pagano et al. 1995). These compounds, for example, inhibit the NADPH oxidase of neutrophils. The lack of effect of inhibitors of xanthine oxidase, cyclooxygenase, and nitric oxide synthase indicate that these are not involved in significant superoxide anion generation in normal tissue (Pagano et al. 1995). Studies on superoxide generating enzyme activity in the rabbit aorta showed that NADPH was the preferred substrate for superoxide anion formation (Pagano et al. 1995). Although this enzymatic activity has not been fully characterized, its major difference from that contained in neutrophils is that it appears to be constitutively active, whereas the neutrophil enzyme requires activation by a protein kinase C dependent pathway. In addition, antibodies which recognize portions of the rabbit neutrophil enzyme do not react with the rabbit aortic proteins, thus distinguishing the vascular source of super-

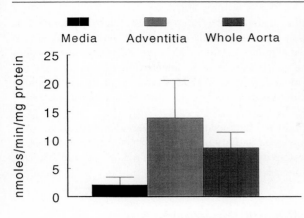

Fig. 2. Superoxide generating activity is localized primarily in the adventitia of the normal rabbit aorta (Pagano et al. 1995). Superoxide anion generating activity was measured in a particulate fraction of homogenates of rabbit aorta using the cytochrome C assay. The activity present in whole aortic fractions is accounted for primarily by that present in the adventitia. (From Pagano et al. 1995)

oxide anion (Pagano et al. 1995). Preliminary characterization indicates that the aortic enzyme may be regulated by intracellular calcium levels (Maridonneau-Parini et al. 1986) and arachidonic acid (Rubinek and Levy 1993) which are two characteristics similar to the neutrophil enzyme.

Of interest is the localization of the majority of the superoxide generating activity in the rabbit aorta to adventitial cells. This was demonstrated both enzymatically on separate fractions of the rabbit aorta (Fig. 2) (Pagano et al. 1995) and in cross sections of the aorta stained with nitroblue tetrazolium which is reduced to blue formazan by superoxide anion (Fig. 3). This may represent superoxide generating activity in fibroblasts, because dermal fibroblasts have previously been described to possess an

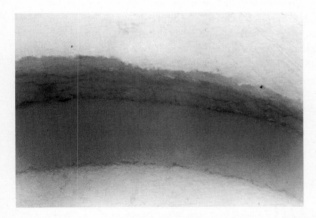

Fig. 3. Photomicrograph of a ring of normal rabbit aorta which was treated with DDC (10 mM) for 30 min and then stained in the presence of nitroblue tetrazolium (100 μm) for 90 min at 37 °C. The nitroblue tetrazolium is reduced to blue formazan by superoxide anion, which is observed as dark staining primarily in the adventitia. Blue staining is present in endothelium as well and could represent generation in these cells of superoxide anion or nitric oxide by which nitroblue tetrazolium is also reduced. The photomicrograph was taken directly from the surface of the paraffin block in which the aortic ring was embedded. Original magnification was ×40

NADPH oxidase activity similar to that in the blood vessel (Meier et al. 1991). The activity is unlikely to be solely associated with sympathetic nerves, as these are confined to the adventitial-medial junction. Other cells in the normal vascular wall as well have been shown to have constitutively active superoxide generating enzymes including a NADH oxidase in endothelial cells (Mohazzab-H et al. 1994), and a NADH/NADPH oxidase in smooth muscle cells (Griendling et al. 1994; Mohazzab-H and Wolin 1994).

At least two mechanisms might be imagined to increase vascular superoxide anion levels. These would include increased activity of superoxide generating enzymes, or alternatively a decrease in the activity of SOD. As discussed below both mechanisms may be important in the pathogenesis of vascular disease.

Interaction of Nitric Oxide and Superoxide Anion

Nitric oxide is recognized to react with superoxide anion. Early studies showed that ambient superoxide levels in physiological buffers decreased the half-life of nitric oxide (Gryglewski et al. 1986; Rubanyi and Vanhoutte 1986). The reaction between nitric oxide and superoxide is rapid, occurs with a similar rate constant as the reaction of super-oxide with SOD, and results in formation of peroxynitrite, a highly reactive oxidant which has been itself recently implicated in vascular disease by its ability to cause lipid peroxidation and protein tyrosine nitrosation (Beckman et al. 1990) (Fig. 4). It is pos-sible that because it is so reactive, that peroxynitrite significantly contributes to cellu-lar dysfunction when superoxide levels are high and nitric oxide is being produced (Beckman et al. 1990). In the absence of nitric oxide, superoxide is converted to hydro-gen peroxide, which by reaction with reduced iron forms hydroxyl radical, which in turn can mediate lipid peroxidation, DNA damage and alterations in gene expression (Ito et al. 1996) (Fig. 4). Thus, when superoxide anion is increased either as a result of decreased scavenging or increased production, formation of several reactive oxygen species may contribute to tissue damage and dysfunction.

The importance of the regulation of superoxide anion to normal nitric oxide func-tion is demonstrated by the fact that endothelium-dependent relaxation is decreased when superoxide dismutase activity is inhibited with DDC (Mugge et al. 1991a). The

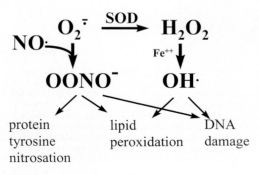

Fig. 4. This schema shows the disposi-tion of superoxide anion to other reac-tive molecules. Superoxide anion is con-verted to highly reactive peroxynitrite by reaction with nitric oxide (NO) or to hydrogen peroxide (H_2O_2) by SOD. H_2O_2 can be further converted to hydroxyl radical (OH) by ferrous ion (Fe^{++}). The more reactive species are thought to mediate deleterious cellular effects

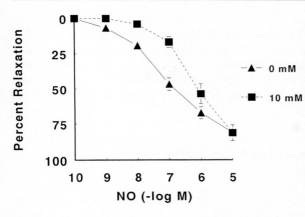

Fig. 5. Superoxide anion arising as a result of inhibiting SOD with DDC inhibits relaxations to nitric oxide. Data show relaxations of normal rabbit thoracic aortic rings under control conditions or after treatment for 30 min with DDC (10 mM). Rings were contracted with phenylephrine to 40%–50% of maximal contractile tone and then exposed to increasing concentrations of nitric oxide gas dissolved in water. The relaxations are expressed as a percentage of maximal relaxations to sodium nitroprusside

effect of inhibiting SOD can also be demonstrated by measuring the direct response of the smooth muscle to nitric oxide itself (Fig. 5). Relaxations to low concentrations of nitric oxide (10^{-9}–10^{-7} M) are greatly inhibited when SOD activity is blocked by treatment with DDC. Higher concentrations of nitric oxide are not as greatly affected, and at 10^{-5} M, nitric oxide relaxes normally despite elevated levels of superoxide anion. This may be related to the fact that at higher concentrations, nitric oxide decreases superoxide levels as shown in Fig. 6 (Pagano et al. 1993). Measurements of lucigenin

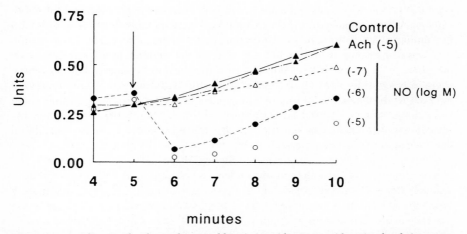

Fig. 6. Superoxide anion levels are decreased by nitric oxide. Superoxide anion levels in groups of individual rabbit aortic rings were measured from 4 to 10 min following treatment with DDC (10 mM) during additions of acetylcholine (Ach, 10^{-5} M), or nitric oxide (10^{-7}–10^{-5} M). Nitric oxide decreased levels of superoxide anion, whereas acetylcholine had no effect indicating that insufficient quantities of nitric oxide are released by the endothelium-dependent agent to decrease superoxide anion levels. This is either due to the fact that the concentrations of nitric oxide released from the endothelium are too low, or that the release from the endothelium is unable to effectively decrease superoxide anion arising in the adventitia. (From Pagano et al. 1993)

chemiluminescence show that concentrations of nitric oxide greater than 10^{-7} M decrease superoxide levels. As shown, acetylcholine does not affect these greatly elevated levels of superoxide anion in the rabbit aorta. This probably explains why endothelium-dependent relaxation to acetylcholine is inhibited, whereas relaxations to higher concentrations of exogenously added nitric oxide persist after treatment with DDC. These experiments demonstrate the importance of normal SOD activity in normal nitric oxide function.

Arachidonic Acid, Vasoactive Eicosanoids, and Superoxide Anion

When endothelial cells are activated, the calcium rise activates not only nitric oxide synthase, but also phospholipase A_2 which liberates arachidonic acid from membrane phospholipids. A portion of the arachidonic acid is metabolized to eicosanoids, some of which are highly vasoactive. Prostacyclin formed as a result of cyclooxygenase activity vasodilates some, but not all vasculature. Under some conditions, particularly in blood vessels in which prostacyclin does not vasodilate, the eicosanoids formed produce vasoconstriction. This difference likely results from the lack of mechanisms which cause adenylate cyclase stimulation by prostacyclin, and/or a greater sensitivity of smooth muscle PGH_2/thromboxane A_2 receptors. For instance, in the normal rabbit aorta arachidonic acid causes vasoconstriction which depends on conversion in the endothelial cells to a vasoconstrictor identified as PGH_2 (Pagano et al. 1991). PGH_2, the product of cyclooxygenase, is the precursor of all other prostaglandins and thromboxane A_2. As a consequence, it is the prostanoid produced in greatest amounts, and it has similar contractile potency as thromboxane A_2. In the rabbit aorta, the formation of a vasoconstrictor from arachidonic acid in the endothelium is prevented by inhibitors of cyclooxygenase, but not of thromboxane synthase, and has been chemically identified as PGH_2 (Pagano et al. 1991). The contractions are also prevented by PGH_2/thromboxane A_2 receptor antagonists. These studies suggest that when arachidonic acid is released endogenously in sufficient quantities, and when the smooth muscle is adequately sensitive to PGH_2/thromboxane A_2 receptor stimulation, that the major vasoactive product is PGH_2.

Studies of the effects of exogenously added PGH_2 on rabbit aorta show that it is a potent inhibitor of endothelium-dependent relaxation and that this inhibition uniquely depends on superoxide anion (Tesfamariam and Cohen 1992a). Figure 7 shows the effect of PGH_2 on endothelium-dependent relaxation to acetylcholine. In the presence of a concentration of the prostanoid which causes minimal contraction on its own, the relaxations to acetylcholine are greatly inhibited and resemble the response in a number of vascular disease states. Although the effect of PGH_2 could be prevented by blocking PGH_2/thromboxane A_2 receptors, the effect was not mimicked by U46619, a thromboxane A_2 mimetic. Of note is the fact that this effect of PGH_2 was highly dependent on superoxide anion, being reversed by exogenously added SOD (Fig. 8). Indeed, contractions to higher concentrations of PGH_2, unlike those to other vasoconstrictors tested such as U46619, were inhibited by SOD as well (Tesfamariam and

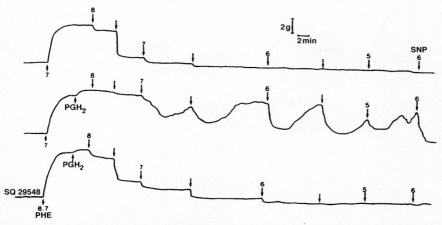

Fig. 7. Exogenous PGH$_2$ inhibits endothelium-dependent relaxation to acetylcholine in the normal rabbit aorta. Recordings of tension are shown of three rings of aorta. The normal relaxation to acetylcholine *(top panel)* is contrasted to a ring in which a threshold concentration of PGH$_2$ (10^{-6} M) is added prior to acetylcholine *(middle panel)*, resulting in impaired relaxation. In the third ring which was treated with the thromboxane A$_2$/PGH$_2$ receptor antagonist, SQ29548 (10^{-6} M), relaxations were normal. PHE, phenylephrine.; SNP sodium nitroprusside. *Numbers* with *arrows* refer to the negative logarithm of the concentration added to the ring. (From Tesfamariam and Cohen 1992a)

Cohen 1992a). This suggests that a unique property of PGH$_2$, perhaps its synthesis to another prostaglandin or other product, explains the vasoconstriction and inhibition of endothelium-dependent relaxation. Although the mechanism has not been identified, it is apparent that the process depends on the generation of superoxide anion and that the superoxide anion decreases the bioactivity of nitric oxide to account for the reduced acetylcholine relaxations. This mechanism is likely more than simple functional antagonism between the vasoconstrictor and nitric oxide, because of the fact that U46619 had no effect. These results were applicable to both the rat (Tesfamariam 1994) and rabbit (Tesfamariam and Cohen 1992a) aorta suggesting that they can be generalized.

Vasoactive Prostanoids and Superoxide Anion in Vascular Disease

There is growing evidence that the factors discussed above play a role in the abnormal vascular reactivity in vascular disease. Decreased scavenging of superoxide anion by SOD and increased destruction of nitric oxide have been implicated in hypercholesterolemia (Mugge et al. 1991; Ohara et al. 1993), hypertension (Nakazono et al. 1991; Sharma et al. 1992; Tschudi et al. 1996), and diabetes (Tesfamariam and Cohen 1992; Cohen 1993). In addition, increased superoxide generation may occur. Under

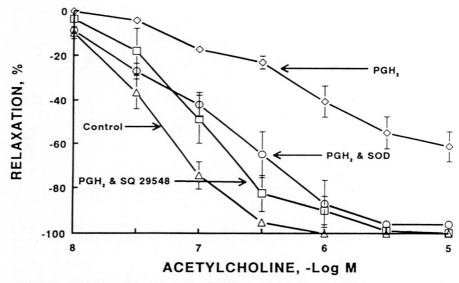

Fig. 8. Data (means ± sem) summarizing experiments such as shown in Fig. 7 illustrate the effect of SOD (150 units/ml) or SQ29548 (10^{-6} M) on the abnormal endothelium-dependent relaxations caused by PGH$_2$. The inhibitory effect of PGH$_2$ on acetylcholine-induced relaxation is prevented by either SOD or by the thromboxane A$_2$/PGH$_2$ receptor antagonist. (From Cohen 1993)

disease conditions, other sources of superoxide anion such as xanthine oxidase in hypertension (Nakazono et al. 1991) and hypercholesterolemia (Ohara et al. 1993), and aldose reductase in diabetes mellitus (Cohen 1993) may increase their contribution to enzymatic sources present in normal tissues.

Hypertension

In the spontaneously hypertensive rat, intravenous injection of a chimeric SOD molecule which binds to extracellular matrix caused hypotension, whereas it had no effect on blood pressure of normotensive rats (Nakazono et al. 1991). The effect of the synthetic SOD was mimicked by oxypurinol, suggesting that xanthine oxidase is the source of superoxide which contributes to hypertension in this hypertensive rat. Recently, the abnormal endothelium-dependent relaxations in spontaneously hypertensive rats were shown to be associated with decreased nitric oxide levels, increased superoxide anion levels, and were normalized by SOD (Tschudi et al. 1996). This suggests that destruction of nitric oxide by superoxide anion may contribute to the abnormal vascular reactivity and the increased blood pressure in hypertension.

Recently, it was demonstrated that hypertension associated with angiotensin II infusion increased the production of superoxide anion by isolated aorta of normal rats (Rajagopalan et al. 1996). The increased superoxide anion production was implicated in the impaired endothelium-dependent relaxation by showing that the reduced relaxations were normalized by treating the blood vessels with SOD. The effect of angiotensin II was prevented by angiotensin II receptor blockade and was not mimicked by hypertension caused by norepinephrine infusion. This suggests that hypertension specifically associated with activation of the renin-angiotensin system could be accompanied by increased superoxide production. The results in the spontaneously hypertensive rat suggest that other systems may also be implicated because this rat model is associated with low renin activity.

Although no link between superoxide anion and vasoactive eicosanoids has been made in hypertension, a role for vasoconstrictor prostanoids has been implicated in abnormal vascular reactivity in a number of studies of human and animal hypertension. Abnormal endothelium-dependent vasodilator responses of the forearm of hypertensive patients were shown to be ameliorated by indomethacin, suggesting the contribution of a vasoconstrictor prostanoid (Taddei et al. 1993). Animal models of hypertension also demonstrate endothelium-dependent vasodilator responses which can be accounted for by production of vasoconstrictor prostanoids (Auch-Schwelk and Vanhoutte 1992; Lin and Nasjletti 1992; Mombouli and Vanhoutte 1993; Dyer et al. 1994; Ge et al. 1995). In rats with aortic coarctation (Lin and Nasjletti 1992; Lin et al. 1994) or deoxycorticosterone-salt induced hypertension (unpublished observations), elevation of endothelial cell calcium with the ionophore, A23187, results in endothelium-dependent contractions. These are blocked by cyclooxygenase inhibitors or PGH_2/thromboxane A_2 receptor antagonists, but not by thromboxane synthase inhibitors, implicating PGH_2 as the vasoconstrictor being produced.

Diabetes Mellitus

Endothelial cell dysfunction also occurs in diabetes mellitus (for review see Cohen 1993). Impaired cholinergic vasodilator responses have been demonstrated in the forearm of diabetic patients (Johnstone et al. 1993) and in isolated penile erectile tissue (Saenz de Tejada et al. 1989), and both large arteries and microvessels have been shown to be affected in rat and rabbit animal models. In the aorta of the alloxan diabetic rabbit, a model of type 1 diabetes, decreased endothelium-dependent relaxations to acetylcholine occur six weeks after elevation of plasma glucose to approximately 300 mg/dl. Because the abnormality could be mimicked by exposure of the isolated aorta of normal rabbits to high concentrations of glucose for 6 hours, the defect in the diabetic animal may be attributed to exposure of the blood vessel to glucose rather than to other metabolic abnormalities associated with diabetes. The abnormal endothelium-dependent relaxations in the aorta were prevented by treating either the diabetic rabbit in vivo or the isolated aorta in vitro with aldose reductase inhibitors. This suggests that this glucose metabolic enzyme which is present in large amounts in endothelial cells and whose activity is increased in diabetes is involved. In addition, the abnormal relaxations of the aorta in both in vivo and in vitro models appear similar to those of the normal aorta exposed to PGH_2 shown in Fig. 9. Furthermore, cyclooxygenase inhibitors and PGH_2/thromboxane A_2 receptor antagonists, but not throm-

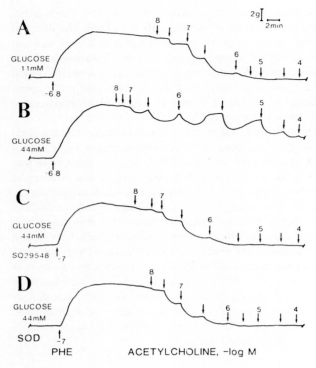

Fig. 9A–D. Recordings show the effect of incubating rabbit aortic rings in elevated glucose solution (44 mM) for 6 h. Compared with the control ring (**A**), the relaxation in the ring exposed to elevated glucose (**B**) is impaired. Pretreatment of the ring with either SQ29548 (10^{-6} M, **C**) or SOD (150 units/ml, **D**) prevented the effect of elevated glucose. (From Tesfamariam et al. 1990)

boxane synthase inhibitors prevented or reversed the abnormal endothelium-dependent relaxations implicating PGH_2 in the abnormal responses in diabetes (Fig. 9). Furthermore, similar to the effects of exogenous PGH_2, the abnormal relaxations in these diabetic models were reversed or prevented by antioxidants, including SOD (Tesfamariam and Cohen 1992b). These studies therefore suggest that superoxide anion and vasoconstrictor prostanoids can contribute to abnormal endothelial function in diabetes.

Conclusions

A number of enzymatic processes give rise to superoxide anion which is normally disposed of by antioxidant enzymes. If the activity of these scavenging systems is impaired as described in a number of vascular diseases, or if the production of superoxide anion is increased, then more reactive and damaging agents including hydrogen peroxide, hydroxyl radical, and peroxynitrite are formed. These reactive oxygen species contribute to vascular dysfunction and may play a role in chronic structural changes in vascular disease.

A role for vasoconstrictor prostanoids in hypertension and diabetes is not universally observed. This likely indicates that the processes leading to the production and response to the prostanoids are not an essential feature of the pathogenesis of these diseases. Rather, their involvement likely reflects a process in a particular vascular bed which results in sufficient release of arachidonic acid, as well as sufficient PGH_2/thromboxane A_2 receptors to mediate vasoconstriction. It is also probably true that oxygen-derived free radicals are not primary mediators of hypertension or diabetes, but rather reflect changes in vascular tissue metabolism which are likely responding to the disease state. These changes may include acute changes in free radical generation resulting from increased metabolism of glucose in diabetes, and hemodynamic forces in hypertension. In addition, blood vessels subjected to either of these diseases undergo chronic inflammatory, fibrotic, and atherosclerotic changes, all of which could be accompanied by, as well as be exacerbated by oxygen-derived free radicals and eicosanoids.

The studies reviewed here provide evidence that oxygen-derived free radicals and vasoconstrictor prostanoids contribute to the mechanisms of reduced nitric oxide-mediated endothelium-dependent relaxations in hypertension and diabetes. Current research in a number of laboratories is directed toward determining the mechanisms by which these abnormalities of vascular function are precipitated by the disease process.

References

1. Abrahamsson T, Brandt U, Marklund SL, Sjoquist PO (1992) Vascular bound recombinant extracellular superoxide dismutase type C protects against the detrimental effects of superoxide radicals on endothelium-dependent arterial relaxation. Circ Res 70:264–271
2. Auch-Schwelk W, Vanhoutte PM (1992) Contractions to endothelin in normotensive and spontaneously hypertensive rats: Role of endothelium and prostaglandins. Blood Pressure 1:45–49
3. Beckman JS, Beckman TW, Chen JC, Marshall PA, Freeman BA (1990) Apparent hydroxyl radical production by peroxynitrite: Implications of endothelial injury from nitric oxide and superoxide. Proc Natl Acad Sci 87:1620–1624
4. Carlsson LM, Jonsson J, Edlund T, Marklund SL (1995) Mice lacking extracellular superoxide dismutase are more sensitive to hyperoxia. Proc Natl Acad Sci 92:6264–6268
5. Cohen RA (1993) Dysfunction of vascular endothelium in diabetes mellitus. Circ 87:V67–V76
6. Cohen RA (1995) The role of nitric oxide and other endothelium-derived vasoactive substances in vascular disease. Prog Cardiovasc Dis 38:105–128
7. Dyer SM, Taylor DA, Bexid S, Hime NJ, Frewin DB, Head RJ (1994) Identification of a nonendothelial cell thromboxane-like constrictor response and its interaction with the renin-angiotensin system in the aorta of spontaneously hypertensive rats. J Vasc Res 31:52–60
8. Faulkner K, Fridovich I (1993) Luminol and lucigenin as detectors for O2. Free Rad Biol Med 15:447–451
9. Fridovich I (1978) The biology of oxygen radicals. The superoxide radical is an agent of oxygen toxicity; superoxide dismutases provide an important defense. Science 201:875–880
10. Ge T, Hughes H, Junquero DC, Wu KK, Vanhoutte PM, Boulanger CM (1995) Endothelium-dependent contractions are associated with both augmented expression of prostaglandin H synthase-1 and hypersensitivity to prostaglandin H2 in the SHR aorta. Circ Res 76:1003–1010
11. Griendling KK, Minieri CA, Ollerenshaw JD, Alexander RW (1994) Angiotensin II stimulates NADH and NADPH oxidase activity in cultured vascular smooth muscle cells. Circ Res 74:1141–1148
12. Gryglewski RJ, Palmer RMJ, Moncada S (1986) Superoxide anion is involved in the breakdown of endothelium-derived vascular relaxing factor. Nature 320:454–456

13. Ito Y, Pagano PJ, Tornheim K, Brecher P, Cohen RA (1996) Oxidative stress increases glyceraldehyde-3-phosphate dehydrogenase mRNA levels in isolated rabbit aorta. Am J Physiol 270:H81–H87
14. Johnstone MT, Creager SJ, Scales KM, Cusco JA, Lee BK, Creager MA (1993) Impaired endothelium-dependent vasodilation in patients with insulindependent diabetes mellitus. Circ 88:2510–2516
15. Lin L, Balazy M, Pagano PJ, Nasjletti A (1994) Expression of prostaglandin H2-ediated mechanism of vascular contraction in hypertensive rats. Circ Res 74:197–205
16. Lin L, Nasjletti A (1992) Prostanoid-mediated vascular contraction in normotensive and hypertensive rats. Eur J Pharmacol 220:49–53
17. Maridonneau-Parini I, Tringale SM, Tauber AI (1986) Identification of distinct activation pathways of the human neutrophil NADPH oxidase. J Immunol 137:2925–2929
18. Marklund SF (1992) Regulation by cytokines of extracellular superoxide dismutase and other superoxide dismutase isoenzymes in fibroblasts. J Biol Chem 267(10):6696–6701
19. Meier B, Cross AR, Hancock JT, Kaup FJ, Jones OT (1991) Identification of a superoxide-generating NADPH oxidase system in human fibroblasts. Biochem J 275:241–245
20. Mohazzab-H KM, Kaminski PM, Wolin MS (1994) NADH oxidoreductase is a major source of superoxide anion in bovine coronary artery endothelium. Am J Physiol 266:H2568–H2572
21. Mohazzab-H KM, Wolin MS (1994) Properties of a superoxide anion-generating microsomal NADH moxidoreductase, a potential pulmonary artery pO2 sensor. Am J Physiol 267:L823–L831
22. Mombouli JV, Vanhoutte PM (1993) Purinergic endothelium-dependent and -independent contractions in rat aorta. Hypertension 22:577–583
23. Mugge A, Elwell JH, Peterson TE, Harrison DG (1991a) Release of intact endothelium-derived relaxing factor depends on endothelial superoxide dismutase activity. Am J Physiol 260:C219–C225
24. Mugge A, Elwell JH, Peterson TE, Hofmeyer TG, Heistad DD, Harrison DG (1991b) Chronic treatment with polyethylene-glycolated superoxide dismutase partially restores endothelium-dependent vascular relaxations in cholesterol-fed rabbits. Circ Res 69:1293–1300
25. Nakazono K, Watanabe N, Matsuno J, Sasaki J, Sato T, Inoue M (1991) Does superoxide underlie the pathogenesis of hypertension? Proc Natl Acad Sci 88:10045–10048
26. Ohara Y, Peterson TE, Harrison DG (1993) Hypercholesterolemia increases endothelial SO production. J Clin Invest 91:2546–2551
27. Pagano PJ, Lin L, Sessa WC, Nasjletti A (1991) Arachidonic acid elicits endothelium-dependent release from the rabbit aorta of a constrictor prostanoid resembling prostaglandin endoperoxides. Circ Res 69:396–405
28. Pagano PJ, Tornheim K, Cohen RA (1993) Superoxide anion production by the rabbit thoracic aorta: Effect of endothelium-derived nitric oxide. Am J Physiol 265:H707–H712
29. Pagano PJ, Ito Y, Tornheim K, Gallop P, Cohen RA (1995) An NADPH oxidase superoxide generating system in the rabbit aorta. Am J Physiol 268:H2274–H2280
30. Rajagopalan S, Kurz S, Munzel T, Tarpey M, Freeman BA, Griendling KK, Harrison DG (1996) Angiotensin II-mediated hypertension in the rat increases vascular superoxide production via membrane NADH/NADPH oxidase activation. J Clin Invest 97:1916–1923
31. Rubanyi GM, Vanhoutte PM (1986) Superoxide anions and hyperoxia inactivate endothelium-derived relaxing factor. Am J Physiol 250:H822–H827
32. Rubinek T, Levy R (1993) Arachidonic acid increases the activity of the assembled NADPH oxidase in cytoplasmic membranes and endosomes. Biochim Biophys Acta 1176:51–58
33. Saenz de Tejada I, Goldstein I, Azadzoi K, Krane RJ, Cohen RA (1989) Impaired neurogenic and endothelium-dependent relaxation of human penile smooth muscle: the pathophysiological basis for impotence in diabetes mellitus. NEJM 320:1025–1030
34. Sharma RC, Crawford DW, Kramsch DM, Sevanian A, Jiao Q (1992) Immunolocalization of native antioxidant scavenger enzymes in early hypertensive and atherosclerotic arteries. Arterioscler Thromb 12:403–415
35. Stratlin P, Karlsson K, Johansson BO, Marklund SL (1995) The interstitium of the human arterial wall contains very large amounts of extracellular superoxide dismutase. Arterioscler Thromb Vasc Biol 15:2032–2036
36. Taddei S, Virdis A, Mattei P, Salvetti A (1993) Vasodilation to acetylcholine in primary and secondary forms of human hypertension. Hypertension 21:929–933
37. Tesfamariam B, Brown ML, Deykin D, Cohen RA (1990) Elevated glucose promotes generation of endothelium-derived vasoconstrictor prostanoids in rabbit aorta. J Clin Invest 85:929–932

38. Tesfamariam B (1994) Selective impairment of endothelium-dependent relaxations by prostaglandin endoperoxide. J Hypertens 12:41–47
39. Tesfamariam B, Cohen RA (1992a) Role of superoxide anion and endothelium in vasoconstrictor action of prostaglandin endoperoxide. Am J Physiol 31:H1915–H1919
40. Tesfamariam B, Cohen RA (1992b) Free radicals mediate endothelial cell dysfunction caused by elevated glucose. Am J Physiol 263:H321–H326
41. Tschudi MR, Mesaros S, Luscher TF, Malinski T (1996) Direct in situ measurement of nitric oxide in mesenteric resistance arteries. Hypertension 27:32–35

Endothelial Factors and Myocardial Function

A. M. SHAH

Introduction

The heart contains two types of endothelial cells, namely the vascular endothelial cells lining the coronary vessels and the endocardial endothelial cells lining the inner surfaces of the cardiac chambers. It is now recognised that these cardiac endothelial cells influence myocardial function through the paracrine release of a variety of diffusible substances (Shah 1996). The initial suggestion that endothelial cells influence myocardial function came from Brutsaert et al. (1988) who reported that selective denudation of the endocardial endothelium in isolated cardiac papillary muscle preparations altered contractile behaviour: the duration of twitch contraction was abbreviated and peak force slightly reduced. It was subsequently shown that endocardial endothelial cells modified the inotropic response of isolated papillary muscle preparations to various receptor-mediated agonists, analogous to endothelium-dependent regulation of vascular tone. In bioassay studies using cultured ventricular endocardial endothelial cells and isolated papillary muscle preparations, Shah and colleagues (Smith et al. 1991) demonstrated that these effects were mediated by the release of diffusible factors by endocardial cells. The physiological relevance of these in vitro findings has been controversial given that the endocardial endothelial monolayer in the whole heart is in close contact with only a tiny proportion of the total myocardial mass. In contrast, coronary microvascular endothelial cells are intimately apposed to cardiac myocytes throughout the heart, so that a physiologically relevant paracrine interaction is more plausible. Indeed, recent studies from a number of laboratories have confirmed the existence of such an interaction both in vitro and in humans in vivo.

Cardioactive agents released by endothelial cells include nitric oxide, endothelin-1, prostanoids, and other substances so far characterised only in bioassay studies. Cardiac endothelial cells can also alter local levels of substances such as angiotensin II and bradykinin secondary to their angiotensin converting enzyme (ACE)/kininase activities. "Endothelial" factors such as endothelin-1 and nitric oxide may be expressed in cardiac myocytes themselves, often under pathological situations, potentially comprising a parallel autocrine pathway. The present review focuses on the paracrine effects of endothelial mediators on myocardial contractile function, and the potential physiological and pathophysiological roles of this pathway. Constraints of space prevent the citation of all original references, but these may be found in the reviews cited in this manuscript.

Nitric Oxide and Cardiac Function

The pivotal inter- and intra-cellular signalling role of nitric oxide in many physiological processes is now well recognised. Its molecular biology and biochemistry have been reviewed in detail (Schmidt et al. 1993). Endothelium-derived nitric oxide may indirectly influence myocardial function secondary to alterations in (a) arterial tone and cardiac load, (b) venous return and capacitance, and thus ventricular filling volume, and (c) coronary vascular tone and coronary perfusion. In addition to these indirect actions, nitric oxide exerts *direct* effects on cardiac myocytes. The precise nature of these effects may depend on several factors, such as the cellular source of nitric oxide (e.g., endothelial cell or cardiac myocyte), the amount released (isoform-dependent), physiological variables (e.g., heart rate, coronary flow rate, cardiac loading, level of adrenergic and cholinergic activation), interaction with other paracrine factors, and the presence or absence of pathology (e.g., immune activation or inflammation). The subcellular actions of nitric oxide in myocardium are also potentially complex, probably involving both cGMP-dependent and -independent pathways. Activation of the former could in theory result in activation of cGMP-dependent protein kinase (PKG), activation or inhibition of phosphodiesterases, activation of phosphatases, and/or activation of cGMP-gated ion channels. Possible cGMP-independent effects of nitric oxide include a reduction in oxygen consumption (Shen et al. 1994), alterations in mitochondrial function, and modulation of ion channels. Cellular sources of nitric oxide within the heart include endothelial cells, cardiac myocytes (Schulz et al. 1992), nerve tissue (Klimaschewski et al. 1992) and white blood cells (Wildhirt et al. 1995). The nitric oxide synthase (NOS) isoform expressed constitutively in endothelial cells is NOS-3 (or endothelial NOS, eNOS).

Paracrine Effect of Nitric Oxide on Basal Systolic and Diastolic Function

Experimental data from studies in several different preparations and species indicate that the characteristic effect of endothelium-derived nitric oxide on basal myocardial function is to enhance myocardial relaxation, reduce diastolic tone and slightly reduce peak contractile performance. These actions appear to be mediated through elevation of myocardial cGMP and activation of PKG. In isolated papillary muscle preparations (e.g., ferret, cat), stimulation of nitric oxide release from endocardial endothelium by a specific agonist, substance P, caused earlier isometric twitch relaxation and a slight reduction in peak force, without any change in maximal rate of force development (Smith et al. 1991; Mohan et al. 1995; Fig. 1). These effects were associated with elevation of cGMP and were abolished by denudation of endocardial endothelium or by haemoglobin, which inactivates nitric oxide. Similar effects were observed with the exogenous nitric oxide donor, sodium nitroprusside, or with a lipid-soluble cGMP analogue, 8-bromo-cGMP (Shah et al. 1990; Smith et al. 1991). In isolated ejecting ("working") guinea-pig hearts, studied under constant filling pressure, afterload and heart rate, both sodium nitroprusside and substance P induced a dose-dependent earlier onset and acceleration of left ventricular (LV) pressure fall (Grocott-Mason et al.

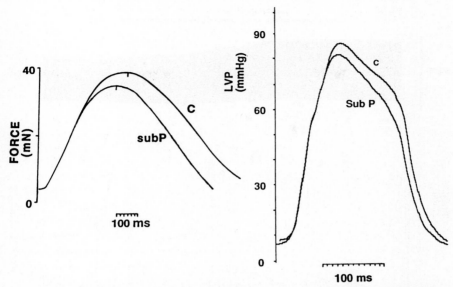

Fig. 1. Representative examples of the effect of endothelium-derived nitric oxide on basal myocardial function. *Left:* Isometric twitch contraction of isolated ferret papillary muscle following stimulation by substance P (*sub* P, 1 µM). *Right:* High-fidelity left ventricular pressure *(LVP)* traces recorded from isolated ejecting guinea-pig heart before and after exposure to Sub P (0.1 µM). c, control trace

1994a,b) without significant changes in LV dP/dt$_{max}$, a pattern similar to that observed in isolated papillary muscles (Fig. 1). This LV relaxant effect was not attributable to the increase in coronary flow induced by these agents, and was consistent with a direct action of nitric oxide on cardiac myocytes in the whole heart. Interestingly, similar acute effects on LV function were recently reported with the ACE inhibitor, captopril, in this preparation (Anning et al. 1995). Captopril induced faster LV relaxation that was inhibited by the bradykinin B$_2$ receptor antagonist HOE140 or by haemoglobin, suggesting an involvement of an endogenous bradykinin/nitric oxide pathway in the response.

In isolated single rat ventricular myocytes, 8-bromo-cGMP (used as a surrogate for nitric oxide) induced a similar pattern of effect to that described above, i.e., an earlier onset of isotonic twitch relaxation and a reduction in myocyte twitch shortening without significant reduction in shortening velocity (Shah et al. 1994b). These changes were not associated with changes in the cytosolic Ca^{2+} transient, indicating that they were mediated by a cGMP-induced reduction in myofilament response to Ca^{2+} (Fig. 2). Consistent with this mechanism, steady-state tetanic shortening of intact cardiac myocytes (achieved by high frequency stimulation following inhibition of the sarcoplasmic reticulum ATPase by thapsigargin) was inhibited by 8-bromo-cGMP without concomitant reduction in steady-state peak tetanic Ca^{2+}. 8-bromo-cGMP effects were inhibited by KT5823, a specific inhibitor of PKG. A previous study in skinned porcine

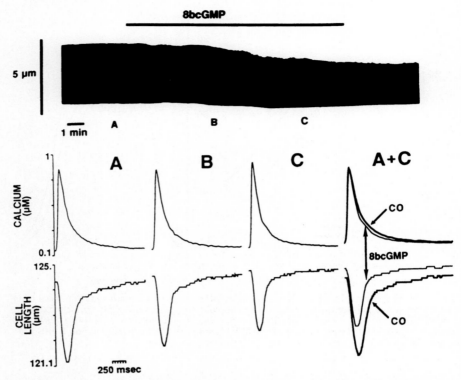

Fig. 2. Effect of 8-bromo-cGMP (*8bcGMP*, 50 μM) on an isolated rat ventricular myocyte loaded with the fluorescent probe indo-1 free acid. *Top:* Chart recording of myocyte shortening. *Below:* Expanded scale cytosolic Ca^{2+} transients and twitch contractions at baseline *(A)*, 4 min *(B)* and 11 min *(C)* after 8bcGMP addition. (From Shah et al. 1994b)

cardiac fibres also reported a PKG-induced reduction in the Ca^{2+} sensitivity of force production (Pfitzer et al. 1982). In addition to the changes in twitch shortening and relaxation in rat myocytes, 8-bromo-cGMP significantly prolonged diastolic length without any change in diastolic Ca^{2+} (Shah et al. 1994b). This acute increase in diastolic length, also observed in quiescent (electrically unstimulated) cardiac myocytes, may reflect a reduction in intrinsic diastolic tone in these externally mechanically-unloaded cells. It implies that active cross-bridge cycling during diastole may contribute to "diastolic tone" and that this can be modulated by changes in Ca^{2+}-myofilament interaction.

Recent studies have explored the presence of similar effects of nitric oxide in human subjects in vivo (Paulus et al. 1994, 1995). The effect of low-dose bicoronary sodium nitroprusside infusion on basal LV function was studied in patients with normal coronary arteries and normal LV function being investigated for atypical chest pain. Sodium nitroprusside infusion resulted in a significant reduction in peak LV systolic pressure and an earlier onset of LV relaxation (a reduced time to LV dP/dt_{min}) without any change in LV dP/dt_{max} (Fig. 3), effects similar to those observed in the in vitro

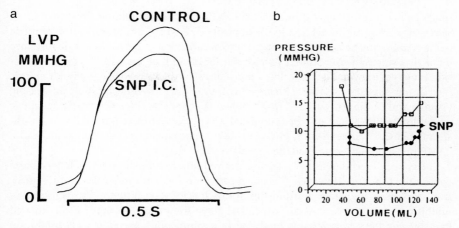

Fig. 3a,b. Representative example of effect of bicoronary sodium nitroprusside infusion (*SNP I.C.*; 4 µg/min for 5 min, divided between the two arteries) on high-fidelity left ventricular pressure (*LVP*, a) and diastolic pressure-volume relations (b) in a normal human subject undergoing diagnostic cardiac catheterization. (From Shah et al. 1995a)

studies (see above). In addition, LV end-diastolic pressure was reduced despite an increase in LV end-diastolic volume, with a downward shift in the end-diastolic pressure-volume relation, consistent with increased end-diastolic distensibility (Paulus et al. 1994). No reduction in LV ejection fraction or stroke volume was observed. A systemic effect of sodium nitroprusside was unlikely since infusion of an identical dose into the right atrium failed to reproduce these effects. In subsequent studies using an identical protocol, bicoronary substance P infusion caused similar effects on LV function both in subjects with normal coronaries and LV function and in cardiac transplant recipients free of rejection or graft vasculopathy being catheterised at annual follow-up (Paulus et al. 1995). Right atrial infusion of substance P again failed to reproduce these effects. These studies provide the first evidence for a paracrine effect of coronary endothelial factors (probably nitric oxide) on human LV contractile function.

Nitric Oxide and Beta-Adrenergic Inotropic Response

A number of studies have provided evidence for an interaction between nitric oxide and beta-adrenergic inotropic response. In adult rat ventricular myocytes, inhibition of endogenous nitric oxide by L-arginine analogues augmented the positive inotropic response to isoproterenol (Balligand et al. 1993, 1995). In human patients with LV dysfunction (Hare et al. 1995), the increment in LV dP/dt_{max} after intracoronary dobutamine was greater in the presence of NO synthase inhibitors. These studies imply that nitric oxide (probably generated within cardiac myocytes) reduces beta-adrenergic inotropic response. In human cardiac transplant recipients, coronary endothelium-derived nitric oxide had considerably greater negative inotropic activity in the presence of concurrent beta-adrenergic stimulation with dobutamine (Bartunek et al. 1995), suggesting an interaction in the opposite direction – i.e., modification of nitric oxide response by beta-agonists rather than modification of beta-adrenergic response by nitric oxide. Whether similar interactions occur in humans with normal LV function is not yet known. The underlying mechanism of the nitric oxide–beta-adrenergic interaction appears to involve changes in L-type Ca^{2+} current. In adult guinea-pig (Wahler and Dollinger 1995) and frog (Mery et al. 1994) myocytes, the exogenous nitric oxide donor, SIN-1, had no effect on basal L-type Ca^{2+} current but inhibited cAMP-stimulated Ca^{2+} current via a PKG-dependent mechanism in the guinea-pig and a cAMP-phosphodiesterase-mediated mechanism in the frog.

It has been considered that myocyte-derived nitric oxide does not influence basal contractile function. However, a recent study in rat myocytes reported that the positive shortening-frequency relationship in this species was shifted upwards following inhibition of NOS (Kaye et al. 1996). The force-frequency relationship in isolated hamster papillary muscles was modified in a similar manner after NOS inhibition (Finkel et al. 1995). The subcellular mechanism of these effects remains to be elucidated.

Role of Nitric Oxide in Parasympathetic Responses

In neonatal rat myocytes, the negative chronotropic action of muscarinic agonists was blocked by NOS inhibitors, suggesting that nitric oxide mediated this response (Balligand et al. 1993). In spontaneously beating adult rabbit sino-atrial cells (Han et al. 1994) and atrioventricular nodal cells (Han et al. 1996), cholinergic inhibition of isoproterenol-stimulated Ca^{2+} current was mediated via intra-myocyte generation of nitric oxide. In anaesthetised dogs, vagal inhibition of the inotropic response to dobutamine was attenuated by the NOS inhibitor, L-NMMA, suggesting that at least part of the effect was mediated by nitric oxide (Hare et al. 1995), although the cell type responsible was not defined. However, in frog myocytes (Mery et al. 1996) and adult guinea-pig ventricular myocytes (Stein et al. 1993) the response to cholinergic agonists did not involve nitric oxide. It appears therefore that there may be significant species differences in these responses.

Positive Inotropic Effects of Nitric Oxide

Recent studies indicate the potential for a positive inotropic action of nitric oxide, particularly at low doses. In isolated guinea-pig (Wahler and Dollinger 1995) and frog (Mery et al. 1994) myocytes, low concentrations of SIN-1 increased cAMP-stimulated Ca^{2+} current, possibly via inhibition of a cAMP-phosphodiesterase and a consequent increase in cAMP levels. In human atrial myocytes, low doses of SIN-1 (>1 pM) markedly increased basal Ca^{2+} current, via a cGMP-inhibited phosphodiesterase (Kirstein et al. 1995). In isolated adult rat myocytes, 8-bromo-cGMP induced a transient positive inotropic effect associated with an increase in cytosolic Ca^{2+} in 50% of cells

(Shah et al. 1994b). In cat papillary muscle (Mohan et al. 1996) and adult rat myocytes (Kojda et al. 1996), low concentrations of nitric oxide donors caused positive inotropic effects whereas higher concentrations were negatively inotropic.

Physiological Roles of NOS-3(eNOS)-Derived Nitric Oxide

The enhancement of LV relaxation and increase in end-diastolic distensibility induced by endothelium-derived nitric oxide may facilitate both ventricular filling and sub-endocardial coronary perfusion, by increasing the diastolic interval and the driving pressure for cardiac filling and by reducing extravascular compressive forces. Since the most relevant physiological stimuli for nitric oxide release from coronary endothelium in vivo are probably flow-induced shear stress and cyclical mechanical deformation during the cardiac cycle (Lamontagne et al. 1992), the above effects could be particularly important in situations where heart rate and pulsatile coronary flow are increased (e.g., exercise). An endogenous inotropic mechanism that is highly dependent on the "diastolic function" of the heart is the Frank-Starling response, i.e., a length (or volume)-dependent increase in myocardial contractile function. Recent studies in isolated ejecting hearts suggest that endogenous nitric oxide augments the Frank-Starling response, possibly secondary to an increase in diastolic chamber distensibility (Prendergast et al. 1997b). As discussed above, nitric oxide also appears to influence the myocardial force-frequency relationship. Myocyte-derived nitric oxide has a significant role in modulating beta-adrenergic inotropic responses and, at least in some species, in mediating cholinergic effects on heart rate. The potential relevance of these nitric oxide-mediated responses across a broad range of mammalian species, and particularly in humans, remains to be established. Studies to date have investigated only acute influences on myocardial function. Whether chronic changes in nitric oxide pathways, either physiological or pathophysiological, can influence cardiac structure and function is an important area for investigation.

Pathophysiological Roles of NOS-3(eNOS)-Derived Nitric Oxide

Vascular endothelial dysfunction is well recognised in conditions such as ischaemia-reperfusion, hypertension, LV hypertrophy, diabetes, and heart failure. What role, if any, altered nitric oxide production plays in the myocardial dysfunction of these conditions remains to be established. In a number of situations, exogenous nitric oxide/cGMP appear to exert beneficial actions on myocardial function. In an experimental model of acute "diastolic dysfunction", induced by brief hypoxia-reoxygenation of isolated single rat cardiac myocytes, pretreatment with 8-bromo-cGMP completely inhibited abnormalities of relaxation and diastolic tone (Shah et al. 1995b). In analogous experiments in isolated rat hearts, pretreatment with sodium nitroprusside reduced abnormal LV relaxation following a 5 min period of hypoxia, independent of changes in coronary flow (Draper and Shah 1997). It is possible that nitric oxide may be similarly beneficial in other cardiac conditions characterised by "diastolic dysfunction". In a model of myocardial stunning following brief (10 min) coronary occlusion in conscious dogs, NOS inhibition enhanced myocardial depression independent of changes in coronary flow, implying that nitric oxide had a protective effect (Hasebe et al. 1993). Nitric oxide also protected isolated rat myocytes against reoxygenation injury following prolonged anoxia (Schlüter et al. 1994), in part by effects on osmotic fragility (Schlüter et al. 1996). However, nitric oxide also has the potential to exert deleterious effects on myocardium, e.g., through reaction with superoxide during periods of oxidant stress to form toxic peroxynitrite radicals (Beckman et al. 1990).

NOS-2 (iNOS) and Pathophysiology

NOS-2 (or inducible NOS) is expressed in the heart in several conditions, e.g., endotoxic shock, cardiac allograft rejection, myocarditis and myocardial infarction. Cardiac cell types shown to express NOS-2 include cardiomyocytes, endothelial cells and white blood cells. In vitro studies have shown that expression of NOS-2 is generally associated with depression of myocardial contractile function. However, the situation in vivo may be more complex, since NOS-2-derived nitric oxide could exert beneficial effects such as anti-microbial and anti-viral activity, coronary vasodilatation, antiplatelet activity, decreased leucocyte adhesion, and enhancement of LV relaxation. The potential pathophysiological role of NOS-2 has been discussed in detail elsewhere (Morris and Billiar 1994; Shah 1996).

Endothelin

The endothelin family of peptides (Yanagisawa et al. 1988) is distributed widely in many tissues and has numerous biological actions (reviewed by Rubanyi and Polokoff 1994). Endothelin-1 is the only isoform secreted by endothelial cells, and is the most potent vasoconstrictor identified to date. Plasma levels of endothelin-1 are in the femtomolar range and it probably acts as a paracrine factor; its release from vascular endothelial cells occurs preferentially towards the abluminal surface. Both cultured cardiac microvascular endothelial cells (Nishida et al. 1993) and endocardial endothelial cells (Mebazaa et al. 1993) release endothelin-1. Cardiac myocytes can themselves produce the peptide, e.g., during hypoxia or ischaemia (Suzuki et al. 1993; Tonnessen et al. 1995).

Exogenous Endothelin-1 and Myocardial Function

Both ET_A and ET_B endothelin receptors are present on mammalian (including human) cardiac myocytes, and are coupled to multiple signalling pathways including stimulation of phosphoinositide hydrolysis, activation of arachidonic acid metabolism, and G protein-mediated inhibition of adenylyl cyclase. In vitro studies have consistently demonstrated a potent, prolonged positive inotropic action of exogenous endothelin-1 in the subnanomolar concentration range, although there is inter-species and -tissue variation with respect to the precise nature of effect, e.g., the effect on myocardial relaxation. The positive inotropic response probably involves both an elevation of cyto-solic Ca^{2+} and an increase in myofilament responsiveness to Ca^{2+}, mediated at least in part via protein kinase C-induced activation of Na^+-H^+ exchange (Kramer et al. 1990). The relative contributions of these mechanisms to the inotropic activity may vary depending, for example, on the underlying state of myofilament activation (Wang and Morgan 1992). Endothelin-1 exerts negative chronotropic effects through ET_A receptors, particularly during beta-adrenergic stimulation, while activation of ET_B receptors may have opposing actions (Ono et al. 1995).

Role of Endogenous Endothelin in the Heart

Despite these data regarding the myocardial activity of exogenous endothelin-1, whether endogenous endothelin-1 has a physiological role in regulating cardiac func-tion has been uncertain. Endothelin-1 concentrations that exert myocardial effects could potentially induce ischaemia secondary to coronary vasoconstriction, thereby counteracting the positive inotropic activity. Recent studies have provided evidence in support of a physiological role for endogenous endothelin-1. Cultured ventricular endocardial endothelial cells were shown to tonically release sufficient endothelin-1 to augment the shortening of isolated cardiac myocytes (Mebazaa et al. 1993). In isolated rat hearts, endothelin-1 concentrations in the coronary circulation were adequate to exert positive inotropic effects on isolated trabeculae (McClellan et al. 1994). In isola-ted ferret papillary muscles, exposure to the ET_A antagonist, BQ123, resulted in a pro-gressive (over about 60 min) abbreviation of isometric twitch with a minor reduction in peak force; BQ123 had no effect in muscles denuded of endocardial endothelium (Evans et al. 1994; Fig. 4). These data are consistent with reversal of a tonic effect of endothelin-1, released from endocardial endothelial cells in situ, to prolong twitch duration and delay relaxation. The relatively slow effect of BQ123 can be accounted for by the nature of the endothelin-receptor interaction and recycling process (Marsault et al. 1993). Similar data were reported in isolated rat trabeculae where exposure to BQ123 caused progressive reduction in peak isometric force, consistent with reversal of basal positive inotropic activity of endothelin-1 in this preparation (McClellan et al. 1995). Recent data indicate that BQ123 induces progressively earlier and faster LV relaxation in isolated buffer-perfused guinea-pig hearts (Prendergast et al. 1997a). Whether similar effects of endogenous endothelin-1 are relevant in vivo remains to be established.

Endothelin-1 released endogenously within the heart may interact with other endo-thelial autacoids. For example, endothelin-1 induces the release of nitric oxide from endothelial cells and ANP from atrial tissue, while both nitric oxide and ANP inhibit endothelin-1 production by endothelial cells (Rubanyi and Polokoff 1994). Angiotensin II stimulates the expression of endothelin-1 mRNA in rat endothelial cells and cultured rat myocytes.

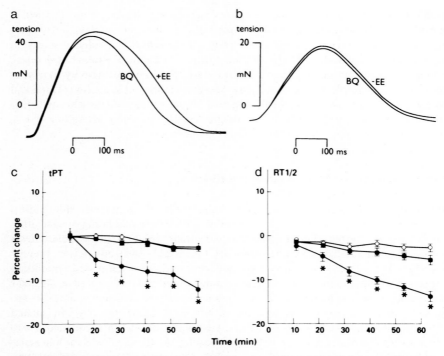

Fig. 4a–d. Effect of ET_A receptor antagonist BQ123 (10 µM) on isolated ferret papillary muscle preparations. **a,b** Effect on isometric force in endocardial endothelium-intact (+*EE*) and endocardial endothelium-denuded preparation (–*EE*). **c,d** Effect on time to peak tension (*tPT*) and time to 50% tension decline (*RT1/2*). *Open circles,* untreated muscles (time controls); *filled circles,* BQ123 +EE group; *filled squares,* BQ123 -EE group. * $p < 0.05$ for comparison between groups by ANOVA

Endothelin and Myocardial Pathophysiology

Increasing evidence supports a pathophysiological role of endothelin-1. Cardiac endothelin-1 production is increased during myocardial ischaemia, with a significant source being ischaemic cardiac myocytes themselves (Tonnessen et al. 1995). Binding of endothelin-1 to cardiac receptors is increased following prolonged ischaemia and during reoxygenation. An increase in endothelin-1 could be beneficial by improving contractile performance or deleterious by causing coronary vasoconstriction, increasing myocardial oxygen demand, and impairing relaxation. Administration of BQ123 or antiendothelin monoclonal antibodies reduced infarct size in several experimental studies (Grover et al. 1993). Endothelin-1 is known to induce hypertrophy of neonatal rat cardiac myocytes, and recent studies implicate endogenous endothelin-1 also in angiotensin II-induced hypertrophy (Ito et al. 1993). An endothelin antagonist was reported to reduce pressure overload-induced hypertrophy in rats in vivo (Ito et al. 1994).

Novel Low-Molecular-Weight Modulators of Cardiac Myofilaments

Recent experimental findings suggest that endothelial cells release other, as yet chemically unidentified, substances that influence myocardial contraction by altering contractile protein properties. Bioassay studies performed using cultured endocardial or vascular endothelial cells as the donor tissue and isolated single rat cardiac myocytes as the assay tissue demonstrated the tonic release into superfusing buffer of a potent negative inotropic substance, termed "myofilament desensitizing agent" (Shah et al. 1994a; Pepper et al. 1995). Endothelial cell superfusate with this activity caused a rapid and reversible reduction in shortening of isolated myocytes, an earlier onset of relaxation, and an increase in diastolic length (Fig. 5). These effects were not accompanied by significant changes in the cytosolic Ca^{2+} transient, indicating that they resulted from reduction in the myofilament response to Ca^{2+}. This mechanism was confirmed in intact single myocytes that were tetanically stimulated, where tetanic shortening was reduced without reduction in tetanic Ca^{2+}. The effects were not blocked by inhibition of nitric oxide synthesis, nitric oxide action (with haemoglobin), cyclooxygenase, adenosine receptors, cGMP-dependent protein kinase, cAMP-dependent protein kinase, protein kinase C, or pertussis toxin-sensitive G-proteins, and were not associated with changes in cytosolic pH. The "myofilament desensitizing" activity was retained at 37 °C for several hours, was not destroyed by proteases, and was found in low molecular weight (<1 kDa) fractions (Shah et al. 1994a).

Ramaciotti et al. (1993) used a different approach in which the coronary effluent of isolated buffer-perfused rat hearts was tested for its effects on isolated rat cardiac trabeculae. They reported the presence in coronary effluent of both positive and negative inotropic factors presumed to be released by endothelial cells, since these activities were absent if endothelial cells in the hearts were damaged. The positive inotropic factor was shown to be endothelin-1, but the identity of the negative inotropic factor is unknown. The factor reduced actin-activated actomyosin ATPase activity in cryostat sections of myocardial tissue. It was suggested that release of this factor was regulated by coronary flow rate. We have recently studied the effects of rat heart coronary effluent with negative inotropic activity on contraction of isolated myocytes. As with "myofilament desensitizing agent", coronary effluent also reduced myocyte shortening without reducing cytosolic Ca^{2+} transients (Pepper et al. 1995), suggesting that these activities may be the same. The physiological and/or pathophysiological relevance of this factor remains yet to be defined.

Under conditions of moderate hypoxia (PO2 40–50 torr) cultured endothelial cells release into superfusing buffer a different substance clearly distinguishable from "myofilament desensitizing agent". Hypoxic endothelial superfusate induces potent reversible inhibition of myocyte contraction accompanied by a reduction in diastolic myocyte length, independent of changes in cytosolic Ca^{2+} (Shah et al. 1997). These actions involve direct inhibition of actomyosin ATPase and crossbridge cycling, independent of subcellular signalling pathways. This factor could potentially be important in pathophysiological conditions involving myocardial hypoxia or ischaemia, e.g., myocardial hibernation.

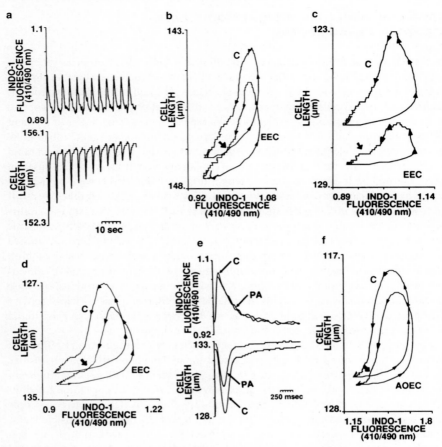

Fig. 5a–f. Effect of cultured endothelial cell superfusates with "myofilament desensitizing activity" on contraction and indo-1 fluorescence ratio (Ca^{2+} transient) of isolated rat myocytes. *EEC*, endocardial endothelial cells; *PA*, pulmonary artery endothelial cells; *AOEC*, aortic endothelial cells; *C*, baseline control trace in each case. **a** Fast speed chart recording showing rapid onset of negative inotropic action. **b–d, f** Phase-plane loops of instantaneous Ca^{2+} versus length during twitch contractions demonstrating a right and downward shift of relaxation trajectory *(arrows)*, consistent with a reduction in relative myofilament response to Ca^{2+}. **e** Superimposed Ca^{2+} transients and twitch contractions indicating a negative inotropic effect without significant reduction in Ca^{2+} transient. (From Shah et al. 1994a)

Other Agents

Cardiac endothelial regulation of myocardial function may potentially involve several other factors. Endothelial cells release natriuretic peptides (Suga et al. 1992; Brandt et al. 1995), and natriuretic peptides can modulate myocardial contractile function. Natriuretic peptides also influence the release of endothelial autacoids such as endothelin-1, while endothelial factors in turn may modulate release of natriuretic peptides by cardiac myocytes. Both coronary vascular and cultured endocardial endothelial cells release prostanoids, e.g., prostacyclin and PGE2, particularly during hypoxia (Mebazaa et al. 1995). The direct effects of prostanoids on myocardial function remain uncertain, but it is suggested that they may modulate inotropic responses to other agonists (Mohan et al. 1995). Adenylpurines (e.g., ATP, adenosine) are released from endothelial cells especially during hypoxia or ischaemia, and may affect myocardial contractile function.

Summary and Conclusions

The data reviewed in this article strongly support the existence of a paracrine pathway for the regulation of myocardial contractile function by endothelial cells. Mediators that are implicated include nitric oxide, endothelin-1, and other less well characterised substances. Of interest is the fact that many of these mediators act predominantly by modulating cardiac myofilament responsiveness to Ca^{2+}, a subcellular mechanism that distinguishes them from most conventional inotropic regulatory factors, e.g., catecholamines. This mode of action often results in a disproportionate effect on myocardial relaxation and diastolic tone, with the potential to influence cardiac filling and coronary perfusion. Nitric oxide and the "myofilament desensitizing agent" tend to enhance relaxation and reduce diastolic tone whereas endothelin-1 may have opposing effects. The endothelial pathway is likely to act in concert and to interact with other cardio-

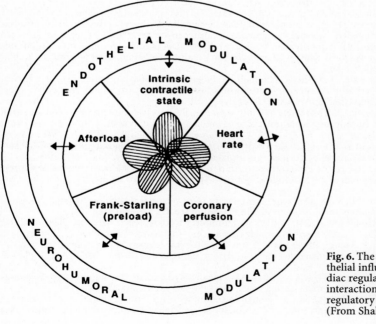

Fig. 6. The role of endothelial influences in cardiac regulation and their interaction with other regulatory pathways. (From Shah 1994)

vascular regulatory pathways, e.g., neurohumoral influences, heart rate, loading, and coronary perfusion (Fig. 6). Important factors regulating cardiac endothelial function include flow-induced shear, mechanical forces, ambient PO_2, and many receptor-dependent agonists. Under pathological conditions, there may be alterations both in the release of endothelial factors and in the responses of diseased myocardium to these factors. Intra-myocyte production of "endothelial" mediators (e.g., nitric oxide, endothelin) is implicated in some disease states. The role of the endothelial pathway in conditions characterised by diastolic dysfunction and/or endothelial dysfunction merits attention. A better understanding of the role of this pathway in the physiological regulation of heart structure and function and in the pathophysiology of disease states may lead to novel therapeutic strategies.

Acknowledgements. This work was supported by the UK Medical Research Council and the British Heart Foundation. Dr Shah is supported by an MRC Senior Clinical Fellowship. Thanks are due to all the collaborators in the studies referred to in this manuscript.

References

Anning PB, Grocott-Mason RM, Lewis MJ, Shah AM (1995) Enhancement of left ventricular relaxation in the isolated heart by an angiotensin converting-enzyme inhibitor. Circulation 92:2660–2665

Balligand JL, Kelly RA, Marsden PA, Smith TW, Michel T (1993) Control of cardiac muscle function by an endogenous nitric oxide signaling system. Proc Natl Acad Sci USA 90:347–51

Balligand JL, Kobzik L, Han X, Kaye DM, Belhassen L, O'Hara DS, Kelly RA, Smith TW, Michel T (1995) Nitric oxide-dependent parasympathetic signaling is due to activation of constitutive endothelial (type III) nitric oxide synthase in cardiac myocytes. J Biol Chem 270:14582–14586

Bartunek J, Paulus WJ, Vanderheyden M, Shah AM (1995) Dobutamine modifies myocardial contractile response to receptor-mediated coronary endothelial stimulation (abstract). Circulation 92 (suppl I):I–790

Beckman JS, Beckman TW, Chen J, Marshall PA, Freeman BA (1990) Apparent hydroxyl radical production by peroxynitrite: implications for endothelial injury from nitric oxide and superoxide. Proc Natl Acad Sci USA 87:1620–1624

Brandt RR, Heublein DM, Mattingly MT, Pittelkow MR, Burnett JC (1995) Presence and secretion of atrial natriuretic peptide from cultured human aortic endothelial cells. Am J Physiol 268:H921–H925

Brutsaert DL, Meulemans AL, Sipido KR, Sys SU (1988) Effects of damaging the endocardial surface on the mechanical performance of isolated cardiac muscle. Circ Res 62:358–366

Draper NJ, Shah AM (1997) Pretreatment with a nitric oxide donor reduces contractile depression and improves relaxation following brief hypoxia in isolated rat hearts. J Mol Cell Cardiol 29:1195–1205

Evans HG, Lewis MJ, Shah AM (1994) Modulation of myocardial relaxation by basal release of endothelin from endocardial endothelium. Cardiovasc Res 28:1694–1699

Finkel MS, Oddis CV, Mayer OH, Hattler BG, Simmons RL (1995) Nitric oxide synthase inhibitor alters papillary muscle force-frequency relationship. J Pharmacol Exp Ther 272:945–952

Grocott-Mason R, Anning PB, Evans H, Lewis MJ, Shah AM (1994a) Modulation of left ventricular relaxation in isolated ejecting heart by endogenous nitric oxide. Am J Physiol 267:H1804–H1813

Grocott-Mason R, Fort S, Lewis MJ, Shah AM (1994b) Myocardial relaxant effect of exogenous nitric oxide in the isolated ejecting heart. Am J Physiol 266:H1699–H1705

Grover GJ, Dzwonczyk S, Parham CS (1993) The endothelin-1 receptor antagonist BQ123 reduces infarct size in a canine model of coronary occlusion and reperfusion. Cardiovasc Res 27:1613–1618

Han X, Shimoni Y, Giles WR (1994) An obligatory role for nitric oxide in autonomic control of mammalian heart rate. J Physiol 476:309–314

Han X, Kobzik L, Balligand JL, Kelly RA, Smith TW (1996) Nitric oxide synthase (NOS$_3$)-mediated cholinergic modulation of Ca^{2+} current in adult rabbit atrioventricular nodal cells. Circ Res 78:998–1008

Hare JM, Keaney JF, Balligand JL, Loscalzo J, Smith TW, Colucci WS (1995) Role of nitric oxide in parasympathetic modulation of beta-adrenergic myocardial contractility in dogs. J Clin Invest 95:360–366

Hare JM, Loh E, Creager MA, Colucci WS (1995) Nitric oxide inhibits the positive inotropic response to beta-adrenergic stimulation in humans with left ventricular dysfunction. Circulation 92:2198–2203

Hasebe N, Shen YT, Vatner SF (1993) Inhibition of endothelium-derived relaxing factor enhances myocardial stunning in conscious dogs. Circulation 88:2862–2871

Ito H, Hirata Y, Adachi S, Tanaka M, Tsujino M, Koike A, Nogami A, Marumo F, Hiroe M (1993) Endothelin-1 is an autocrine/paracrine factor in mechanism of angiotensin II-induced hypertrophy in cultured rat cardiomyocytes. J Clin Invest 92:398–403

Ito H, Hiroe M, Hirata Y, Fujisaki H, Adachi S, Akimoto H, Ohta Y, Marumo F (1994) Endothelin ETA receptor antagonist blocks cardiac hypertrophy provoked by hemodynamic overload. Circulation 89:2198–2203

Kaye DM, Wiviott SD, Balligand JL, Simmons WW, Smith TW, Kelly RA (1996) Frequency-dependent activation of a constitutive nitric oxide synthase and regulation of contractile function in adult rat ventricular myocytes. Circ Res 78:217–224

Kirstein M, Rivet-Bastide M, Hatem S, Bénardeau A, Mercadier JJ, Fischmeister R (1995) Nitric oxide regulates the calcium current in isolated human atrial myocytes. J Clin Invest 95:794–802

Klimaschewski L, Kummer W, Mayer B, Couraud JY, Preissler U, Philippin B, Heym C (1992) Nitric oxide synthase in cardiac nerve fibers and neurons of rat and guinea pig heart. Circ Res 71:1533–1537

Kojda G, Kottenberg K, Nix P, Schluter KD, Piper HM, Noack E (1996) Low increase in cGMP induced by organic nitrates and nitrovasodilators improves contractile response of rat ventricular myocytes. Circ Res 78:91–101

Kramer BK, Smith TW, Kelly RA (1990) Endothelin and increased contractility in adult rat ventricular myocytes. Role of intracellular alkalosis induced by activation of the protein kinase C-dependent Na+-H+ exchanger. Circ Res 68:269–279

Lamontagne D, Pohl U, Busse R (1992) Mechanical deformation of vessel wall and shear stress determine the basal release of endothelium-derived relaxing factor in the intact rabbit coronary vascular bed. Circ Res 70:123–130

Marsault R, Feolde E, Frelin C (1993) Receptor externalization determines sustained contractile responses to endothelin-1 in the rat aorta. Am J Physiol 264:C687–C693

McClellan G, Weisberg A, Rose D, Winegrad S (1994) Endothelial cell storage and release of endothelin as a cardioregulatory mechanism. Circ Res 70:787–803

McClellan G, Weisberg A, Winegrad S (1995) Endothelin regulation of cardiac contractility in absence of added endothelin. Am J Physiol 268:H1621–H1627

Mebazaa A, Mayoux E, Maeda K, Martin L, Lakatta EG, Robotham JL, Shah AM (1993) Paracrine effects of endocardial endothelial cells on myocyte contraction mediated via endothelin. Am J Physiol 265:H1841–H1846

Mebazaa A, Wetzel R, Cherian M, Abraham M (1995) Comparison between endocardial and great vessel endothelial cells: morphology, growth and prostaglandin release. Am J Physiol 268:H250–H259

Mery PF, Pavoine C, Belhassen L, Pecker F, Fischmesiter R (1994) Nitric oxide regulates cardiac Ca^{2+} current. J Biol Chem 268:26286–26295

Mery PF, Hove-Madsen L, Chesnais JM, Hartzell HC, Fischmesiter R (1996) Nitric oxide synthase does not participate in negative inotropic effect of acetylcholine in frog heart. Am J Physiol 270:H1178–H1180

Mohan P, Brutsaert DL, Sys SU (1995) Myocardial performance is modulated by interaction of cardiac endothelium derived nitric oxide and prostaglandins. Cardiovasc Res 29:637–640

Mohan P, Brutsaert DL, Paulus WJ, Sys SU (1996) Myocardial contractile response to nitric oxide and cGMP. Circulation 93:1223–1229

Morris SM, Billiar TR (1994) New insights into the regulation of inducible nitric oxide synthesis. Am J Physiol 266:E829–E839

Nishida M, Springhorn JP, Kelly RA, Smith TW (1993) Cell-cell signaling between adult rat ventricular myocytes and cardiac microvascular endothelial cells in heterotypic primary culture. J Clin Invest 91:1934–1941

Ono K, Eto K, Sakomoto A, Masaki T, Shibata K, Sada T, Hashimoto K, Tsujimoto G (1995) Negative chronotropic effect of endothelin 1 mediated through ETA receptor in guinea pig atria. Circ Res 76:284–292

Paulus WJ, Vantrimpont PJ, Shah AM (1994) Acute effects of nitric oxide on left ventricular relaxation and diastolic distensibility in man. Circulation 89:2070–2078

Paulus WJ, Vantrimpont PJ, Shah AM (1995) Paracrine coronary endothelial control of left ventricular function in humans. Circulation 92:2119–2162

Pepper CB, Lang D, Lewis MJ, Shah AM (1995) Endothelial inhibition of myofilament calcium response in intact cardiac myocytes. Am J Physiol 269:H1538–H1544

Pfitzer G, RŸegg JC, Flockerzi V, Hofmann F (1982) cGMP-dependent protein kinase decreases calcium sensitivity of skinned cardiac fibers. FEBS Lett 149:171–175

Prendergast BP, Anning PB, Lewis MJ, Shah AM (1997a) Regulation of left ventricular relaxation in the isolated guinea-pig heart by endogenous endothelin. Cardiovasc Res 33:131–138

Prendergast BD, Sagach V, Shah AM (1997b) Basal release of nitric oxide augments the Frank-Starling Response in the isolated heart. Circulation (in press)

Ramaciotti C, Sharkey A, McClellan G, Winegrad S (1992) Endothelial cells regulate cardiac contractility. Proc Natl Acad Sci USA 89:4033–4036

Ramaciotti C, McClellan G, Sharkey A, Rose D, Wiseberg A, Winegrad S (1993) Cardiac endothelial cells modulate contractility of rat heart in response to oxygen tension and coronary flow. Circ Res 72:1044–1064

Rubanyi GM, Polokoff MA (1994) Endothelins: molecular biology, biochemistry, pharmacology, physiology, and pathophysiology. Pharmacol Rev 46:325–415

Schluter KD, Weber M, Schraven E, Piper HM (1994) NO donor SIN-1 protects against reoxygenation-induced cardiomyocyte injury by a dual action. Am J Physiol 267:H1461–H1466

Schluter KD, Jakob G, Ruiz-Meana M, Garcia-Dorado D, Piper HM (1996) Protection of reoxygenated cardiomyocytes against osmotic fragility by NO donors. Am J Physiol (in press)

Schmidt HHHW, Lohmann SM, Walter U (1993) The nitric oxide and cGMP signal transduction system: regulation and mechanism of action. Biochim Biophys Acta 1178:153–175

Schulz R, Nava E, Moncada S (1992) Induction and potential biological relevance of Ca^{2+}-independent nitric oxide synthase in the myocardium. Br J Pharmacol 105:575–580

Shah AM (1994) Modulation of myocardial contractile function by endothelium. In: Proceedings of the International Society for Heart Research European Section Meeting, ed Haunso S, Kjeldsen K, pp 53–60. Monduzzi Editore, Bologna

Shah AM (1996) Paracrine modulation of heart cell function by endothelial cells. Cardiovasc Res 31:847–867

Shah AM, Lewis MJ, Henderson AH (1990) Effects of 8-bromo-cyclic GMP on contraction and on inotropic response of ferret cardiac muscle. J Mol Cell Cardiol 23:55–64

Shah AM, Mebazaa A, Wetzel RC, Lakatta EG (1994a) Novel cardiac myofilament desensitizing factor released by endocardial and vascular endothelial cells. Circulation 89:2492–2497

Shah AM, Mebazaa A, Yang Z-K, Cuda G, Lankford EB, Pepper CB, Sollott SJ, Sellers JR, Robotham JL, Lakatta EG (1997) Inhibition of myocardial crossbridge cycling by hypoxic endothelial cells. A potential mechanism for matching oxygen supply and demand? Circ Res 80:688–698

Shah AM, Spurgeon H, Sollott SJ, Talo A, Lakatta EG (1994b) 8-bromo cyclic GMP reduces the myofilament response to calcium in intact cardiac myocytes. Circ Res 74:970–978

Shah AM, Prendergast B, Grocott-Mason RM, Lewis MJ, Paulus WJ (1995a) The influence of endothelium-derived nitric oxide on myocardial contractile function. Int J Cardiol 50:225–231

Shah AM, Silverman HS, Griffith EJ, Spurgeon HA, Lakatta, EG (1995b) Cyclic GMP prevents delayed relaxation at reoxygenation following brief hypoxia in isolated cardiac myocytes. Am J Physiol 268:H2396–H2404

Shen W, Xu X, Ochoa M, Zhao G, Wolin MS, Hintze TH (1994) Role of nitric oxide in the regulation of oxygen consumption in conscious dogs. Circ Res 75:1086–1095

Smith JA, Shah AM, Lewis MJ (1991) Factors released from the endocardium of the ferret and pig modulate myocardial contraction. J Physiol 439:1–14

Stein B, DrögemŸller A, MŸlsch A, Schmitz W, Scholz H (1993) Ca++-dependent constitutive nitric oxide synthase is not involved in the cyclic GMP-increasing effects of carbachol in ventricular myocytes. J Pharmacol Exp Ther 266:919–925

Suga S, Nakao K, Itoh H, Komatsu Y, Ogawa Y, Hama N, Imura H (1992) Endothelial production of C-type natriuretic peptide and its marked augmentation by transforming growth factor-beta. J Clin Invest 90:1145–1149

Suzuki T, Kumazaki T, Mitsui Y (1993) Endothelin 1 is produced and secreted by neonatal rat cardiac myocytes in vitro. Biochem Biophys Res Commun 191:823–830

Tonnessen T, Giaid A, Saleh D, Naess PA, Yanagisawa M, Christensen G (1995) Increased in vivo expression and production of endothelin-1 by porcine cardiomyocytes subjected to ischemia. Circ Res 76:767–772

Wahler GM, Dollinger S (1995) Nitric oxide donor SIN-1 inhibits mammalian cardiac calcium current through cGMP-dependent protein kinase. Am J Physiol 268:C45–C54

Wang J, Morgan JP (1992) Endothelin reverses the effects of acidosis on the intracellular Ca^{2+} transient and contractility in ferret myocardium. Circ Res 71:631–639

Wildhirt SM, Dudek RR, Suzuki H, Pinto V, Narayan KS, Bing RJ (1995) Immunohistochemistry in the identification of nitric oxide synthase isoenzymes in myocardial infarction. Cardiovasc Res 29:526–531

Yanagisawa M, Kurihara H, Kimura S, Tomobe Y, Kobayashi M, Mitsui Y, Yazaki Y, Goto K, Masaki T (1988) A novel potent vasoconstrictor peptide produced by endothelial cells. Nature 332:411–415

Vascular Biology of the Endothelin System

D. J. WEBB and G. A. GRAY

Background

The identification and characterisation of endothelin in 1988 [1] followed the discovery that an endothelium-derived relaxing factor (EDRF) mediates the vascular relaxation to acetylcholine [2] and the subsequent identification of this factor as nitric oxide [3]. An additional important stimulus to the discovery of endothelin was the recognition that endothelial cells in culture generate and release a polypeptide vasoconstrictor factor into the bathing medium [4–6]. In a tour de force for cardiovascular biology, Yanagisawa, Masaki and colleagues [1] identified and characterised endothelin as a 21 amino acid peptide and the most potent constrictor then known. Subsequent studies have shown that the endothelin isolated from endothelial cells (now known as endo-thelin-1) is one of a family of 21 amino acid isopeptides (Fig. 1), each of which contains two intra-chain disulphide bridges linking paired cysteine amino acid residues and all of which are formed through the same processing pathway (Fig. 2). Their respective precursor peptides share high sequence homology but are encoded by distinct genes [7]. It is also clear [8] that the endothelin isopeptides share remarkable structural similarity to the sarafotoxins, peptides isolated from the venom of the Israeli burro-wing asp, *Atractaspis engaddensis* (Fig. 1). The sarafotoxins and endothelins share receptor binding sites, signalling pathways and pharmacological actions. Indeed, iso-forms of sarafotoxin have proved valuable tools for characterising endothelin recep-tors [9].

Endothelin-1 is the major isopeptide produced by human endothelial cells and is present in the greatest concentration in blood, although only at picomolar concentra-tions. However, because endothelial endothelin-1 release is predominantly abluminal [10], concentrations are likely to be sufficient to activate vascular receptors. Indeed, recent studies using endothelin receptor antagonists suggest that endothelin-1 is released tonically to maintain basal systemic vascular resistance in man [11]. In this way, endothelin-1 may act as the physiological antagonist of nitric oxide, which is also thought to be released in a tonic manner [12]. However, these mediators differ both in release rate and duration of action. While synthesis of nitric oxide can be increased within minutes in response to various stimuli, endothelin synthesis is regulated at transcriptional level with a resultant delay in release [13, 14]. In addition, nitric oxide has a short half-life and its effects can be terminated quickly by cessation of release whereas the evidence to date suggests that endothelin-1 binds, pseudoirreversibly, to

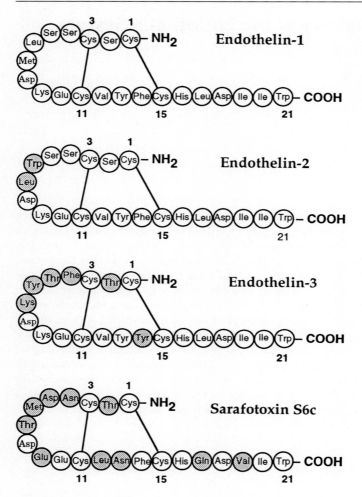

Fig. 1. Structures of the endothelin iso-peptides and sara-fotoxin S6c. The *filled circles* indicate the amino acids that differ from those in ET-1

its receptors on smooth muscle and its constrictor and pressor actions are of long duration [1, 15].

Intensive research carried out since the original description of the endothelin system has greatly improved our appreciation of its physiological significance and its role in cardiovascular disease. Much work has focused on hypertension and this chapter – covering endothelin generation, endothelin receptors and cell signalling pathways – provides a necessary background to that further work.

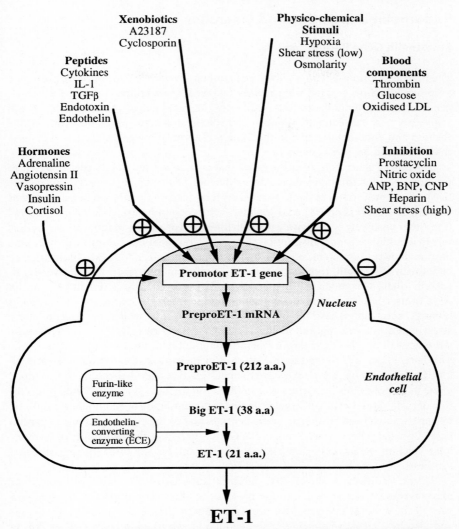

Fig. 2. The generation of endothelin-1 from its precursors and factors influencing this process. *IL-1*, interleukin-1; *TGFβ*, transforming growth factor β; *A23187*, calcium ionophore; *LDL*, low density lipoprotein; *ANP, BNP, CNP*, atrial, brain and c-type natriuretic peptides

Endothelin Generation and Clearance

Endothelin Gene Regulation

The first descriptions of endothelin-1, and cloning of the cDNA encoding its precursor preproendothelin-1 [16, 17], were rapidly followed by identification of at least three genes encoding 'endothelin-like' sequences in mammalian genomes [7]. These sequences were subsequently shown to encode the precursors of endothelin-1, endothelin-2 and endothelin-3 and, in the human genome, these genes are found respectively on chromosomes 6, 1 and 20.

Like other eukaryotic genes, the genes that encode the endothelin precursors have promoter regions through which external factors can modulate transcription. Selective gene modification has demonstrated regions determining the tissue selectivity of endothelin-1 expression [18]. Extracellular factors can influence endothelin-1 generation either positively or negatively through liberation of a series of intracellular mediators that modulate gene transcription (Fig. 2). Several agents, including insulin, thrombin, low density lipoprotein, angiotensin II, vasopressin and endothelin-1 itself enhance endothelin-1 generation via activation of protein kinase C [19]. Responsiveness to protein kinase C is mediated by binding of the proto-oncogenes Jun and Fos to the AP-1 transcription regulatory element of the endothelin-1 promoter [20]. Protein kinase C activation is also thought to be a mechanism by which low levels of shear stress enhance endothelial endothelin-1 release [21]. Interestingly, higher levels of shear stress activate another mechanism that inhibits endothelin-1 mRNA transcription [21]. This latter effect is prevented by inhibitors of nitric oxide synthesis and by methylene blue (an inhibitor of guanylate cyclase), suggesting that endothelial cells release nitric oxide in response to shear stress that inhibits endothelin-1 synthesis through formation of cyclic GMP. Cyclic GMP is also implicated in inhibition of endothelin-1 synthesis by heparin, atrial and brain natriuretic peptides and by the prostanoids prostaglandin E2 and prostacyclin. One action of cyclic GMP is to reduce the availability of intracellular calcium, an action which might be relevant for inhibition of endothelin-1 synthesis.

Recent studies have shown that different isoforms of the endothelin precursor messenger RNA can arise due to the presence of alternative transcription initiation sites in the endothelin-1 gene that are controlled by different promoter regions [18] and also by alternative splicing of the endothelin-2 and endothelin-3 genes during transcription [22]. The consequences for translation efficiency and for subsequent conversion of the endothelin precursors remain to be shown. Future investigations should also demonstrate whether insertion polymorphisms identified in the non-coding region of the endothelin-1 gene [23] might, like polymorphisms in the angiotensin-converting enzyme gene [24], prove to be related to the risk of developing cardiovascular disease.

Endothelin Precursors and Their Processing

All three endothelins are formed through a similar processing pathway from their respective precursor peptides. In the case of endothelin-1 (Fig. 2), removal of the signal sequence on secretion of the 212 amino acid preproendothelin-1 from nucleus to cytoplasm is followed by the first proteolytic step that cleaves proendothelin-1 between Lys^{51}-Arg^{52} and Arg^{92}-Arg^{93} to release the 38 amino acid precursor, big endothelin-1. Furin, a proprotein convertase of the constitutive secretory pathway appears to be the most likely candidate enzyme [25]. Examination of the carboxy terminal portions of proendothelin-1 remaining after cleavage of big endothelin-1 reveals the presence of sequences that encode ,endothelin-like' peptides [13]. Although these peptides have significant structural homology to the mature endothelins, with the relative positions of the four cysteine residues conserved, they do not appear to retain biological activity.

Big endothelin-1 is several orders of magnitude less active than endothelin-1 in displacement of binding to endothelin receptors and in stimulating vasoconstriction [26]. Final processing of big endothelin-1 to release the biologically active 21 amino acid endothelin-1 requires selective cleavage of the Trp^{21}-Val^{22} bond in the carboxy terminal of big endothelin-1, by endothelin-converting enzyme (ECE) activity. Several ECE-like enzyme activities representing different endopeptidase classes have been identified [reviewed in 27]. These include serine proteases, aspartic proteases like pepsin and cathepsin D, and soluble thiol protease. Although inhibitors of these enzymes can prevent conversion of big endothelin-1, their contribution to endothelin-1 biosynthesis is not thought to be of major importance. The physiologically relevant ECE is thought to be a membrane bound, zinc metalloprotease with a narrow neutral pH optimum that is inhibited by the neutral endopeptidase (NEP 24.11) inhibitor, phosphoramidon [28]. The activity of this ECE is not affected by thiorphan or kelatorphan, other NEP inhibitors, or by inhibitors of the neutral metalloprotease angiotensin-converting enzyme. The physiological relevance of the phosphoramidon-sensitive metalloprotease ECE is demonstrated by the ability of phosphoramidon but not thiorphan to inhibit the regional and systemic effects of big endothelin-1 in vivo [29–31].

Following purification and molecular cloning of the human ECE enzyme [32–34], comparison of the deduced amino acid sequences reveals that ECE is a type II integral membrane protein composed of 754/8 amino acids, with a short N-terminal cytoplasmic tail, a hydrophobic transmembrane domain and a large extracellular domain containing a zinc binding motif that is common to the catalytic domains of many metalloproteases. ECE has 10 potential glycosylation sites and expression studies suggest that a high level of glycosylation is important for full enzymatic activity [33]. ECE also has a number of conserved cysteine residues and recent studies have suggested that ECE might exist as a disulphide linked dimer [32]. Amino acid sequences in the C-terminal of the big endothelin isopeptides, particularly residues 27–34, seem to be important determinants of substrate recognition by ECE [35, 36].

The ECE family of metalloprotease enzymes is related to neutral endopeptidase-24.11 (NEP) and the Kell blood group protein but unrelated to angiotensin converting enzyme. ECE-1 [32, 33, 36, 37] appears to be the physiologically active ECE and, by alternative gene splicing [34], it probably exists in at least two different isoforms – ECE-1a and ECE-1b – with functionally distinct roles and tissue distributions [38, 39]. ECE-1a appears to be an intracellular enzyme expressed in the Golgi apparatus of cells, such as the endothelial cells, that synthesise endothelin-1. In contrast, responder cells, such as vascular smooth muscle cells, express extracellular ECE-1b that can convert extracellular big endothelin-1 to mature endothelin-1 [39]. ECE-1 and ECE-2 are both inhibited by phosphoramidon, a combined ECE/NEP inhibitor but not by the selective NEP inhibitor, thiorphan, or by the ACE inhibitor, captopril. Both enzymes are selective for big endothelin-1, raising the possibility that other ECEs with selectivity for big endothelin-2 and -3 will be identified. Through prevention of the generation of the endothelins, ECE-1 must be a potential target for drug treatment.

Endothelin-Converting Enzyme Inhibitors

While phosphoramidon effectively inhibits endothelin-1 formation from big endothelin-1, its therapeutic potential is limited by its low potency and perhaps also by its lack of selectivity for ECE. IC50 values for inhibition of purified ECE by phosphoramidon range from 0.35 μM to 0.8 μM, several orders of magnitude higher than the IC_{50} for inhibition of NEP 24.11 [40, 41]. Several strategies have been used to develop more potent and more selective ECE inhibitors. Compounds based on the structure of phosphoramidon have slightly increased potency and are approximately 400-fold more selective for ECE than phosphoramidon itself, but they retain their relative selectivity for NEP 24.11 [42]. It can be argued that compounds like these and the recently described CGS 26303 [43], which also inhibits NEP 24.11, might have therapeutic advantages over selective ECE inhibitors in that they would potentiate levels of the vasodilator atrial natriuretic peptide, which is broken down by NEP. [D-Val^{22}] Big endothelin-1 (16–38), an analogue of big endothelin-1 inhibits big endothelin-1 conversion in vitro, although its potency is less than that of phosphoramidon [44]. The aspergillomarasmines [45] and more recently the compound WS 79089A (FR 901533) [46] are examples of ECE inhibitors found through screening of fermentation broth. FR 901533 is the only agent so far described that is more selective for ECE than NEP 24.11, while maintaining a similar potency to phosphoramidon for ECE inhibition (IC_{50} = 0.73 μM). The recent cloning of ECE should improve the development of more selective and more potent ECE inhibitors. However, the problem of accessibility will also have to be faced should the intracellular ECE prove to be of more importance physiologically. Intracellular conversion of big endothelin-1 by ECE-1 is inhibited much less potently by phosphoramidon and FR 901533 than extracellular conversion, most likely because of the reduced access of drug to the intracellular site [36].

Sites of Endothelin Synthesis in the Cardiovascular System

Endothelin-1 was first isolated from medium of cultured porcine endothelial cells [1] but is also released by human endothelial cells in culture [47]. In human blood vessels of varying origin, immunoreactive endothelin-1 is detected primarily in association with the endothelial cell layer [48], consistent with the idea that endothelin-1 is released by endothelial cells to act on the underlying smooth muscle cells. *In situ* hybridization for ECE-1 mRNA also shows the most intense labelling over the vascular endothelial cells of most tissues [36]. Removal of the endothelium does not, however, completely prevent the action of big endothelin-1 in the perfused rat mesenteric bed [49], suggesting that smooth muscle cells can also synthesise endothelin-1. Smooth muscle cells in culture express endothelin-1 mRNA and release endothelin-1 [50] but only diffuse immunohistochemical labelling for endothelin-1 can be detected overlying the medial smooth muscle layer of human blood vessels [51]. Interestingly, immuno-histochemical staining for endothelin-1 is intense over atherosclerotic plaques of human coronary arteries [52], particularly over macrophages and intimal smooth muscle cells in active lesions [53]. It may be that only the proliferative phenotype of smooth muscle cells, like those in culture or in atherosclerotic plaques can actively secrete endothelin-1.

To date, the majority of studies that have investigated the presence of the endothelin isopeptides in the human circulation have focused on measurement of immuno-reactive endothelin-1 or its precursor big endothelin-1. However, immunoreactive big endothelin-2, endothelin-3 and big endothelin-3 can also be detected in human plasma [54, 55], suggesting that blood vessels might synthesise other isopeptides in addition to endothelin-1. In human coronary arteries, big endothelin-2 immunoreactivity is detectable in the endothelial cytoplasm [48] and preproendothelin-2 mRNA can be located in the medial layer using in situ hybridisation [22]. There is also evidence that human smooth muscle cells in culture can release endothelin-3 in addition to endo-thelin-1 [56].

Only sparse data is available concerning endothelin formation by cardiac cells. Preproendothelin-1 mRNA is expressed in cultured neonatal rat cardiomyocytes [57] and Northern analysis reveals the presence of ECE-1 mRNA in bovine ventricular tissue [36]. Cultured endocardial cells express endothelin-1 mRNA and also secrete endo-thelin-1 [58] and there is indirect evidence that endothelin-1 is released in vivo by endo-cardial cells lining the ventricular myocardium, particularly following ischaemia [59]. The cellular location of endothelin-1 synthesis in human myocardial tissue remains to be determined.

Clearance and Degradation of the Endothelins

The plasma half-life of endothelin-1 in man is less than 2 min because of its efficient extraction by the splanchnic and renal vascular beds [60, 61]. Endothelin-1 elimina-tion by the lungs may also be important [62]. Extraction of endothelin-1 follows bind-ing to cell surface receptors which are then internalised allowing degradation to be carried out within the cell [63, 64], perhaps in lysosomes [65]. The observation that circulating concentrations of endothelin-1 are increased by a mixed ET_A/ET_B receptor antagonist [66] or by an ET_B receptor selective antagonist [67], but not by ET_A selec-tive antagonists, is suggestive of a role for ET_B type receptors in clearance of endo-thelin-1. Low affinity ET_B type binding sites that might serve this purpose have been found in arteries and veins [68, 69]. A possible candidate for an intracellular degrading enzyme is a soluble protease found in human platelets, vascular smooth muscle and endothelial cells [70, 71]. A deamidase enzyme with similar characteristics was recently purified from rat kidney [72, 73]. The endothelins can also be degraded by NEPs (EC 24.11), which are associated with venous and arterial endothelial cell plasma membranes [74] as well as by an enzyme released from the perfused rat mesenteric bed [75]. Activated polymorphonuclear lymphocyte are able to rapidly inactivate endothelin-1 through release of a protease, believed to be cathepsin G, which degrades endothelin by cleavage of His^{16}-Leu^{17} [76]. This process may have a role in acute inflammation following adhesion of polymorphonuclear lymphocytes to vascular endothelial cells.

Endothelin Receptors

Specific binding sites for [^{125}I] endothelin-1 can be classified according to their relative affinities for the endothelin isopeptides. The ET_A type site is characterised by its very high (subnanomolar) affinity for endothelin-1 and endothelin-2 and its 70 to 100-fold lower affinity for endothelin-3 while the ET_B site has high and equal affinity for all 3 isopeptides (Table 1). These binding characteristics are reflected in the agonist potency of the isopeptides in isolated tissues, demonstrating that the binding sites represent functional receptors [77, 78].

Table 1. Endothelin receptor agonists and antagonists (from [27])

Category	Ligand	Selectivity	Potency (IC_{50} or *, Ki) ET_A	ET_B	Reference
Agonist	ET-1	ET_A/ET_B	160 pM	110 pM	Saeki et al. 1991
	ET-3	ET_B	4.5 nM	70 pM	Saeki et al. 1991
	SRTX S6c	ET_B	4.5 µM	20 pM	Williams et al. 1991
	IRL 1620	ET_B	16 pM	200 pM	Takai et al. 1992
	BQ-3020	ET_B	940 nM	200 pM	Ihara et al. 1992b
Antagonist: peptide	BQ-123	ET_A	7,3 nM	18 µM	Ihara et al. 1991
	FR139317	ET_A	1 nM	7 µM	Aramori et al. 1993
	BQ-485	ET_A	3.4 nM	26 µM	Itoh et al. 1993
	TTA-386	ET_A	0.34 nM	<1 µM	Kitada et al. 1993
	BQ-788	ET_B	1300 nM	1.2 nM	Ishikawa et al. 1994
	Res 701-1	ET_B	>5 µM	10 nM	Tanaka et al. 1994
	PD145065	ET_A/ET_B	3.5 nM	15 nM	Cody et al. 1993
	TAK-044	ET_A/ET_B	0.1 nM	1.8 nM	Kikuchi et al. 1994
Antagonist: non-peptide	97-139	ET_A	1 nM*	1 µM*	Mihara et al. 1994
	BMS 182874	ET_A	55 nM*	>20 µM*	Stein et al. 1994
	PD155080	ET_A	7.4 nM	4.5 µM	Doherty et al. 1995
	Ro 46-8443	ET_B	7 µM	40 mM	Brandli et al. 1996
	Ro 47-0203 (bosentan)	ET_A/ET_B	4.7 nM	95 nM	Clozel et al. 1994
	CGS 27830	ET_A/ET_B	16 nM	295 nM	Mugrage et al. 1993
	SB 209670	ET_A/ET_B	0.2 nM	18 nM	Ohlstein et al. 1994

ET$_A$ and ET$_B$ Receptors: Cloning, Gene Regulation, and Structural Features

Within 2 years of the initial description of endothelin [1] the genes encoding the ET$_A$ and ET$_B$ type receptors that mediate its actions were cloned and characterised [79, 80]. The complementary DNAs (cDNA) encoding the human ET$_A$ and ET$_B$ receptors predict 427 and 442 amino acids respectively and the overall identity between the two mature proteins is reported to be between 55% and 64%, depending on the tissue studied. The ET$_A$ and ET$_B$ receptor genes, located on chromosomes 4 and 13 respectively, have similar structural organisation suggesting that they originated from the same ancestral gene. Although functional studies suggest the existence of further heterogeneity among endothelin receptors (reviewed in [81]), analysis of human genomic DNA with probes specific for the human ET$_A$ and ET$_B$ receptors reveals only two hybridising fragments [82]. Thus, if genes encoding other endothelin receptors exist in the mammalian genome, they must have quite low sequence similarities to the two known endothelin receptor genes. Screening of amphibian cDNA libraries has revealed the existence of two alternative receptor clones, neither of which have yet been detected in the mammalian genome. The first, cloned from *Xenopus* dermal melanophores shows relative selectivity for endothelin-3, consistent with a putative ET$_C$ receptor subtype [83]. The second, cloned from *Xenopus* heart, is termed ET$_{AX}$ because of its high relative affinity for endothelin-1, like the ET$_A$ receptor, but uncharacteristic low affinity for the ET$_A$ receptor selective ligand, BQ-123 [84].

As with the endothelin genes, the non-transcribed 5' flanking regions of the endothelin receptor genes contain a number of regions involved in regulation of gene transcription [85, 86]. Exogenous factors can act through these regions to increase receptor transcription, for example, upregulation of ET$_A$ receptor mRNA by insulin [87] or ET$_B$ receptor mRNA by angiotensin II [88]. These mechanisms may be important in regulation of responsiveness to the endothelins in pathophysiological states. ET$_B$ receptor mRNA is selectively increased in marmosets fed a high cholesterol diet [89] and following glycerol-induced acute renal failure in rats [90]. In contrast, endothelin receptor expression is reduced in atherosclerotic human arteries [91] and in the lungs of rats with pulmonary hypertension [92]. One of the major factors that reduces endothelin receptor number at the cell surface is prolonged exposure to endothelin-1 itself, because of downregulation or feedback inhibition of receptor expression [15], or both of these in combination. In endothelial cells, ET$_B$ receptor expression is decreased because exposure to high local concentrations of endothelin-1 reduces the stability of mRNA molecules rather than reducing transcription [93]. The 3' untranslated regions of both genes contain potential polyadenylation signals that may mediate selective destabilisation of the receptor mRNA [86, 93].

Polymorphisms in the non-coding region of the ET$_A$ receptor gene have been identified in the human genome [23], although the consequences for receptor function are currently unknown. Variants of the endothelin receptor mRNA can arise through alternative splicing of the ET$_A$ and ET$_B$ receptor genes during transcription [94]. The transcribed regions of both receptor genes encode sites for post-translational modification that influence the tertiary structure of the receptor and its linkage to intracellular messenger systems, including consensus sites for N-glycosylation, several potential sites for palmitoylation to anchor the receptor to the cell membrane, and serine residues that may be substrates for regulatory phosphorylation by serine threonine kinases (reviewed in [19]). Phosphorylation may play a role in the downregulation of endothelin receptors that follows prolonged exposure to the endothelin isopeptides. Given the likely importance of the post-translational modification sites for receptor binding and function, it seems possible that some of the receptor heterogeneity observed in functional studies could arise through modification of these sites by alternative splicing. However, at least for the human ET_B receptor, two splice variants do not exhibit any difference in binding or linkage to intracellular signalling pathways [94].

All of the cloned endothelin receptor genes predict a heptahelical membrane spanning structure, common to members of the G-protein-coupled receptor superfamily and similar to many neuropeptide receptors. The regions of greatest sequence conservation between the endothelin receptors and other G protein coupled receptors are concentrated within the hydrophobic transmembrane segments. Amongst the endothelin receptors the 7 transmembrane domains and cytoplasmic loops of the receptors are highly conserved but the N-terminal and other extracellular domains exhibit differences in both length and amino acid sequences [95, 96]. The extracellular N-terminal regions of peptide G-protein-coupled receptors are known to be important for ligand binding. Proteolytic truncation studies reveal that, of the endothelin receptor N-terminal amino acids, only those in closest proximity to the first transmembrane domain are essential for endothelin binding [97]. Computer-assisted molecular modelling has identified Tyr^{129} [98] and Lys^{140} [99] in the second transmembrane domain of the ET_A receptor and ^{181}Lys in the third transmembrane domain of the ET_B receptor [100] as being of potential importance for ligand binding. It has been proposed that the endothelin receptors can be divided into two distinct parts, one comprising transmembrane domains I, II, III and VII that is involved in ligand receptor binding and the other comprising transmembrane domains IV, V and VI that determines isopeptide selectivity [101]. Consistent with this proposal, ET_B-like binding characteristics can be conferred on the ETA receptor by substitution of transmembrane regions IV, V and VI and their intervening loops with the corresponding regions of the ET_B receptor [101, 102]. The predicted third cytoplasmic domain of the endothelin receptors is very short (approximately 30–50 residues), a feature common to G-protein coupled receptors which have peptide ligands [89]. This region and in particular the C-terminal end is implicated in coupling of the human ET_A receptor to G proteins and subsequent liberation of intracellular Ca^{2+} [102, 103].

Agonists at ET_A and ET_B Receptors

All of the endothelin and sarafotoxin peptides possesses four cysteinyl residues that form two disulphide bridges, three polar charged side chains (residues 8–10) and a well conserved hydrophobic C-terminus (residues 16–21) (Fig. 1). Examination of the binding characteristics of these peptides reveals that the ETA receptor has much more rigid structural requirements for ligand binding than the ET_B receptor. Both the amino-terminal loop structure and the carboxy terminal linear portion with Trp in position 21 are vital for high affinity ET_A receptor binding. In contrast, only the linear carboxyl terminal and the Trp[21] is essential for high affinity binding to the ET_B receptor [104]. A number of selective ET_B receptor ligands have been designed based on this linear portion, including BQ-3020 (N-acetyl- [Ala[11,15]] endothelin-1 (6–21) [105] and IRL 1620 (N-succinyl-[Glu[9], Ala[11,15]] endothelin-1 (8–21) [106]. Endothelin-3 and sarafotoxin S6c can also be considered as ET_B selective ligands, endothelin-3 having approximately 2,000-fold and sarafotoxin S6c 30,000-fold selectivity for binding to the ET_B rather than the ET_A receptor [107]. Although both of these ligands contain loop and linear portions like endothelin-1, they have different amino acid sequences within the inner loop portion that might account for their lower affinity at the ET_A receptor.

Endothelin Receptor Antagonists

Since the first description of compounds that could inhibit the binding or actions of endothelin-1 [108, 109], a large number of endothelin receptor antagonists, peptide and non-peptide, selective and non-selective, have become available. The characteristics of some of these are shown in Table 1.

Peptide antagonists have been obtained by chemical modification of endothelin-1 itself, or of microbial products with endothelin receptor binding activity [108, 109]. BQ-123 is a cyclic pentapeptide derived from microbial broth that has relatively high potency for binding to the ET_A receptor subtype [110]. Although several more ET_A antagonists are now available (Table 1), studies using BQ-123 first confirmed the role of endothelin in a number of pathologies [111]. BQ-788 [112] is a peptide compound that is more selective for inhibition of endothelin-1 binding to the ETB receptor (Table 1). The first non-selective endothelin receptor antagonists to be described were also peptides [113, 114]. One of these, TAK-044, is a cyclic hexapeptide with approximately 20-fold higher affinity at the ET_A compared to the ET_B receptor [115]. This compound, unlike many of the other peptide antagonists, has a relatively long duration of action following intravenous administration in vivo [116]. Although useful as research tools the potential of peptides as therapeutic agents may be limited by their short duration of action as well as by their lack of oral availability. With these problems in mind, much research effort has been applied to the development of orally active non-peptide antagonists.

Potent non-peptide antagonists (Table 1) have been developed through optimisation of compounds isolated from plant extracts [117] and microbial broths [118], or screened from chemical libraries [119–121]. These include both ET_A selective agents, such as BMS-182874 [121], and combined $ET_{A/B}$ receptor antagonists, such as bosentan [119]. A number of these drugs are now entering various stages of clinical development in the treatment of cardiovascular disease.

Distribution and Function of Endothelin Receptors in the Cardiovascular System

In vascular tissue, ET_A receptor mRNA is predominantly expressed in smooth muscle [79], while ET_B receptor mRNA is most abundant in endothelial cells [91, 95, 122]. These findings are consistent with the view that constriction of vascular smooth muscle is predominantly mediated by ET_A receptors and that constriction is modified by release of relaxing factors from the endothelium through stimulation of ET_B receptors (Fig. 3). However, in some isolated blood vessels, the constrictor response evoked by endothelin-1 has both ET_A antagonist sensitive and insensitive components (reviewed in [81]) and *in vivo*, pressor responses to endothelin-1 cannot always be completely inhibited by ET_A receptor antagonists [123]. Furthermore, ET_B selective agonists can evoke constriction *in vitro* [124, 125] and pressor responses in vivo [107, 126]. These observations suggest the presence of ET_B receptors that mediate constriction on vascular smooth muscle cells. Indeed, ET_B receptor mRNA is detectable both in the medial smooth

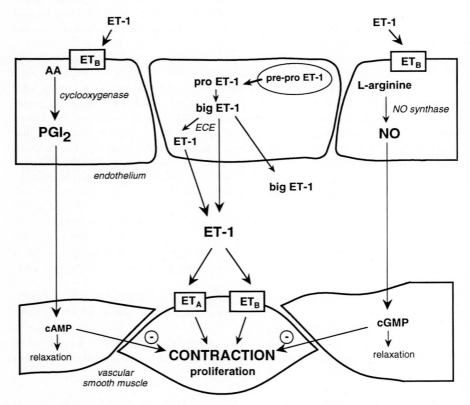

Fig. 3. Influence of the endothelins on vascular tone through effects on vascular smooth muscle cell ET_A receptors, and on vascular smooth muscle and endothelial cell ET_B receptors. (From [27])

muscle of human arteries [127] and in cultured smooth muscle cells [91]. The relative contributions of ET_A and ET_B receptors to vasoconstriction is variable and depends on species and the vessel type studied [128]. ET_B receptors are generally more prevalent in the low pressure venous circulation [129]. In isolated human blood vessels, it is the ET_A receptor subtype that primarily mediates constriction in large calibre arteries [128], but recent studies show that the relative functional role of ET_B receptors is greater in small calibre arteries [130, 131]. The balance of receptors might be altered under pathophysiological conditions. For example, ET_B receptor expression is increased during the change of cultured vascular smooth muscle cells from a contractile to a synthetic phenotype [132], in hypertension [133] and under the influence of angiotensin II [88]. Interestingly, the non-selective endothelin receptor antagonists bosentan [134] and PD 142893 [135] have differential potency on the endothelial ET_B and smooth muscle ET_B receptors suggesting that they might represent discrete receptor subtypes. Subtypes of vascular ET_A receptor that are insensitive to BQ-123 have also been proposed [111, 136].

Although ET_A receptor mRNA cannot be detected in endothelial cells derived from the peripheral vasculature [91, 127], it is reported to be expressed by cultured cerebromicrovascular endothelial cells [137]. High affinity binding sites typical of ET_A receptors can also be detected on rat brain microvascular endothelial cells [138]. These receptors are linked to activation of phospholipase C activation but their functional role in situ is currently unclear.

In the human heart, ET_A and ET_B receptor mRNAs have similar distribution, with both found within the atrioventricular node, the penetrating and branching bundles of His, in atrial and ventricular myocardium, and in endocardial cells [139]. Endothelin-1 is a positive inotropic agent in human atrial and ventricular cardiac muscle strips [140]. In isolated rat cardiomyocytes, endothelin-1 is also reported to have positive chronotropic actions [141]. In addition, within the ventricular myocardium endothelin-1 may have a role in mediating cardiac hypertrophy, as it potentiates the hypertrophic actions of angiotensin II in rat cardiomyocytes [57].

References

1. Yanagisawa M, Kurihara H, Kimura S, Tomobe Y, Kobayashi M, Mitsui Y, Yazaki Y, Goto K, Masaki T (1988) A novel potent vasoconstrictor peptide produced by vascular endothelial cells. Nature 332:411–415
2. Furchgott R, Zawadski J (1980) The obligatory role of endothelial cells in relaxation of arterial smooth muscle to acetylcholine. Nature 288:373–376
3. Palmer R, Ferrige A, Moncada S (1987) Nitric oxide accounts for the biological activity of endothelium-derived relaxing factor. Nature 327:524–526
4. O'Brien RF, McMurtry IF (1984) Endothelial cell supernatants contract bovine pulmonary artery rings. Am Rev Resp Dis 129:A337
5. Hickey KA, Rubanyi G, Paul R, Highsmith R (1985) Characterization of a coronary vasoconstrictor produced by cultured endothelial cells. Am J Physiol 248:C550-C556
6. Gillespie M, Owasoyo J, McMurty I, O'Brien R (1986) Sustained coronary vasoconstriction provoked by a peptidergic substance released from endothelial cells in culture. J Pharmacol Exp Ther 236:339–343
7. Inoue A, Yanagisawa M, Kimura S, Kasuya S, Miyauchi T, Goto K, Masaki T (1989) The human endothelin family: three structurally and pharmacologically distinct isopeptides predicted by three separate genes. Proc Natl Acad Sci USA 86:2863–2867

8. Kloog Y, Ambar I, Sokolovky M, Kochva E, Bdolah A, Wollberg Z (1988) Sarafotoxin, a novel vasoconstrictor peptide: phosphoinositide hydrolysis in rat heart and brain. Science 242:268–270

9. Sokolovsky M (1994) Endothelins and sarafotoxins: receptor heterogeneity. Internat J Biochem 26:335–340

10. Wagner OF, Christ G, Wojta T (1992) Polar secretion of endothelin-1 by cultured endothelial cells. J Biol Chem 267:16066–16068

11. Haynes WG, Webb DJ (1994) Contribution of endogenous generation of endothelin-1 to basal vascular tone. Lancet 344:852–854

12. Vallance P, Collier J, Moncada S (1989) Effects of endothelium-derived nitric oxide on peripheral arteriolar tone in man. Lancet ii:997–1000

13. Yanagisawa M, Inoue A, Ishikawa T, Kasuwa Y, Kimura S, Kumagaye SI, Nakajima K, Watanabe T, Sakakibra S, Goto K, Masaki T (1988) Primary structure, synthesis and biological activity of rat endothelin, an endothelium-derived vasoconstrictor peptide. Proc Natl Acad Sci USA 85:6964–6968

14. Boulanger C, Lüscher TF (1990) Release of endothelin from the porcine aorta: inhibition by endothelium-derived nitric oxide. J Clin Invest 85:587–590

15. Hirata Y, Yoshimi H, Takaichi S, Yanagisawa M, Masaki T (1988) Binding and receptor downregulation of a novel vasoconstrictor endothelin in cultured rat vascular smoth muscle cells. FEBS Lett 239:13–17

16. Itoh Y, Yanagisawa M, Ohkubo S, Kimura C, Kosaka T, Inoue A, Ishida N, Mitsui Y, Onda H, Fujino M, Masaki T (1988) Cloning and sequencing of cDNA encoding the precursor of human endothelium-derived constrictor peptide, endothelin: identity of human and porcine endothelin. FEBS Lett 231:440–444

17. Bloch KD, Freidrich SP, Lee M-E, Eddy RL, Shows TB, Quertermous T (1989) Structural organisation and chromosomal assignment of the gene encoding endothelin. J Biol Chem 264:10851–10857

18. Benatti L, Bonecchi L, Cozzi L, Sarmientos P (1993) Two prepro endothelin mRNAs transcribed by alternative promoters. J Clin Invest 91:1149–1156

19. Gray GA, Webb DJ (1995)The molecular biology and pharmacology of the endothelins. Molecular Biology Intelligence Unit Monograph Series, RG Landes Co., Austin, Texas, USA

20. Lee M-E, Temizer D, Clifford J, Quertermous T. Cloning of the GATA-binding protein that regulates endothelin-1 gene expression in endothelial cells. J Biol Chem (1991) 266:16188–16(192

21. Kuchan MJ, Frangos JA (1993) Shear stress regulates endothelin-1 release via protein kinase C and cGMP in cultured endothelial cells. Am J Physiol 264:H150-H156

22. O'Reilly G, Charnock-Jones DS, Morrison JJ, Cameron IT, Davenport AP, Smith SK (1993) Alternatively spliced mRNAs for human endothelin-2 and their tissue distribution. Biochem Biophys Res Commun 193:834–840

23. Stevens PA, Brown MJ (1995) Genetic variability of the ET-1 and the ETA receptor genes in essential hypertension. J Cardiovasc Pharmacol. 26:S9- S12

24. Cambien F, Costerousse O, Tiret L, Poirier O, Lecerf L, Gonzales MF, Evans A, Arveiler D, Cambou JP, Luc G, Rakotovao R, Ducimetre P, Soubrier F, Alhenc-Gelas F (1994) Plasma level and gene polymorphism of angiotensin-converting enzyme in relation to myocardial infarction. Circulation 90:669–676

25. Laporte S, Denault JB, D'Orleans-Juste P, Leduc R (1993) Presence of furin mRNA in cultured bovine endothelial cells and possible involvement of furin in the processing of the endothelin precursor. J Cardiovasc Pharmacol 22 (Suppl 8):S7-S10

26. Hirata K, Matsuda Y, Akita H, Yokoyama M, Fukuzaki H (1990) Myocardial ischaemia induced by endothelin in the intact rabbit: angiographic analysis. Cardiovasc Res 24:879–883

27. Gray GA, Webb DJ (1996) The therapeutic potential of endothelin receptor antagonists in cardiovascular disease. Pharmacol Ther 72:109–148

28. Opgenorth T, Wu Wong JR, Shiosaki K (1992) Endothelin-converting enzymes. FASEB J 6:2653–2659

29. Fukuroda T, Noguchi T, Tsuchida S, Nishikibe M, Ikemoto F, Okada K, Yano M (1990) Inhibition of biological actions of big endothelin-1 by phosphoramidon. Biochem Biophys Res Commun 172:390–395

30. Matsumura Y, Ikegawa R, Hisaki K, Tsukuhara Y, Takaoka M, Morimoto S (1990) Conversion of big-endothelin-1 to endothelin-1 by two types of metalloproteinases erived from porcine aortic endothelial cells. FEBS Lett 272:166–170

31. McMahon EG, Palomo MA, Moore WM (1991) Phosphoramidon blocks the pressor activity of big endothelin[1–39] and lowers blood pressure in spontaneously hypertensive rats. J Cardiovasc Pharmacol 17 (Suppl 7):S29-S33

32. Schmidt M, Kroger B, Jacob E, Seulberger H, Subkowski T, Otter R, Meyer T, Schmalzing G, Hillen H (1994) Molecular characterization of human and bovine endothelin converting enzyme (ECE-1). FEBS Lett 356:238–243

33. Shimada K, Matsushita Y, Wakabayashi K, Takahashi M, Matsubara A, Iijima Y, Tanzawa K (1995) Cloning and functional expression of human endothelin-converting enzyme cDNA. Biochem Biophys Res Commun 207:807–812

34. Yorimitsu K, Moroi K, Inagaki N, Saito T, Masuda Y, Masaki T, Seino S, Kimura S (1995) Cloning and sequencing of a human endothelin converting enzyme in renal adenocarcinoma (ACHN) cells producing endothelin-2. Biochem Biophys Res Commun 208:721–727

35. Okada K, Takada J, Arai Y, Matsuyama K, Yano M (1991) Importance of the C-terminal region of big endothelin-1 for specific conversion by phosphoramidon-sensitive endothelin converting enzyme. Biochem Biophys Res Commun 180:10(19–1023

36. Xu D, Emoto N, Giaid A, Slaughter C, Kaw S, deWit D, Yanagisawa M (1994) ECE-1: a membrane bound metalloprotease that catalyzes the proteolyic activation of big endothelin-1. Cell 78:473–485

37. Ikura T, Sawamura T, Shiraki T, Hosokawa H, Kido T, Hoshikawa H, Shimada K, Tanzawa T, Kobayashi S, Miwa S, Masaki T (1994) cDNA cloning and expression of bovine endothelin converting enzyme. Biochem Biophys Res Commun 203:1417–1422

38. Valdenaire O, Kalina B, Fischli W, Loeffler B (1995)Cloning and characterisation of two human ECE-1 isoforms. Proceedings of the Fourth International Conference on Endothelin, April 23–26, London

39. Warner TD, Elliott JD, Ohlstein EH (1996) California dreamin', bout endothelin: emerging new therapeutics. TIPS 17:177–179

40. Ohnaka K, Takayanagi R, Nishikawa M, Haji M, Nawata H (1993) Purification and characterization of a phsphoramidon sensitive endothelin-converting enzyme in porcine aortic endothelium. J Biol Chem 268:26759–26766

41. Shimada K, Takahashi M, Tanzawa K (1994) Cloning and functional expression of endothelin-converting enzyme from rat endothelial cells. J Biol Chem 269:18275–18278

42. Fukami T, Hayama T, Amano Y, Nakamura Y, Arai Y, Matsuyama K, Yano M, Ishikawa K (1994) Aminophosphonate endothelin-converting enzyme inhbitors: potency enhancing and selectivity improving modifications of phosphoramidon. Bioorg Med Chem Lett 4:1257–1262

43. De Lombaert S, Ghai RD, Jeng AY, Trapani AJ, Webb RL (1994) Pharmacological profile of a non-peptidic dual inhbitor of neutral endopeptidase 24.11 and endothelin-converting enzyme. Biochem Biophys Res Commun 204:407–412

44. Morita A, Nomizu M, Okitsu M, Horie K, Yokogoshi H, Roller PP (1994) D-Val22 containing human big endothelin-1 analog, (D-Val22)Big ET-1(16–38), inhibits the endothelin converting enzyme. FEBS Lett 353:84–88

45. Matsuura A, Okumura H, Asakura R, Ashizawa N, Takahashi M, Kobayashi F, Ashikawa N, Arai K (1993) Pharmacological profiles of aspergillomarasmines as endothelin converting enzyme inhibitors. Jpn J Pharmacol 187–193

46. Tsurumi Y, Ohhata N, Iwamoto T, Shigematsu N, Sakamoto K, Nishikawa M, Kiyoto S, Okuhara M (1994) WS79089A, B and C, new endothelin converting enzyme inhibitors isolated from Streptosporangium roseum. No. 79089.   J Antibiotics 47:619–631

47. Clozel M, Fischli W (1989) Human cultured endothelial cells do secrete endothelin-1. J Cardiovasc Pharmacol 13:S229-S231

48. Howard P, Plumpton C, Davenport A (1992) Anatomical localization and pharmacological activity of mature endothelins and their precursors in human vascular tissue. J Hypertens 10:1379–1386

49. Hisaki K, Matsumura Y, Nishiguchi S, Fujita K, Takaoka M, Morimoto S (1993) Endothelium independent pressor effect of big endothelin-1 and its inhibition by phosphoramidon in rat mesenteric artery. Eur J Pharmacol 241:75–81

50. Resink T, Hahn AW, Scott-Burden T, Powell J, Weber E, Buhler FR (1990) Inducible endothelin mRNA expression and peptide secretion in cultured human vascular smooth muscle cells. Biochem Biophys Res Commun 168:1308–1310

51. Tokunaga O, Fan J, Watanabe T, Kobayashi M, Kumazaki T, Mitsui Y (1992) Endothelin: immunohistologic localization in aorta and biosynthesis by cultured human aortic endothelial cells. Lab Invest 67:210–217

52. Lerman A, Edwards BS, Hallett JW, Heublein DM, Sandberg SM, Burnett JC (1991) Circulating and tissue endothelin immunoreactivity in advanced atherosclerosis. N Engl J Med 325:997–1001
53. Zeiher AM, Goebel H, Schachinger V, Ihling C (1995) Tissue endothelin-1 immunoreactivity in the active coronary atherosclerotic plaque: A clue to the mechanism of increased vasoreactivity of the culprit lesion in unstable angina. Circulation 91:941–947
54. Matsumoto H, Suzuki N, Kitada C, Fujino M (1994) Endothelin family peptides in human plasma and urine: their molecular forms and concentrations. Peptides 15:505–510
55. Gerbes AL, Moller S, Gulberg V, Henriksen JH (1995) Endothelin-1 and -3 plasma concentrations in patients with cirrhosis: Role of splanchnic and renal passage and liver function. Hepatology 21:735–739
56. Yu JCM, Davenport AP (1995) Secretion of endothelin-1 and endothelin-3 by human cultured vascular smooth muscle cells. Br J Pharmacol 114:551–557
57. Ito H, Hirata Y, Adachi S, Tanaka M, Tsujino M, Koike A, Nogami A, Marumo F, Hiroe M (1993) Endothelin-1 is an autocrine/paracrine factor in the mechanism of angiotensin II-induced hypertrophy in cultured rat cardiomyocytes. J Clin Invest 92:398–403
58. Mebazaa A, Mayoux E, Maeda K (1993) Paracrine effect of endocardial endothelial cells on myocyte contraction mediated via endothelin. Am J Physiol 265:H1841–H1846
59. Tonnessen T, Naess PA, Kirkeboen KA, Ilebekk A, Christensen G (1993) Alterations in plasma endothelin during passage through the left heart chambers before and after brief myocardial ischaemia. Cardiovasc Res 27:2160–2163
60. Weitzberg E, Ahlborg G, Lundberg JM (1991) Long-lasting vasoconstriction and efficient regional extraction of endothelin-1 in human splanchnic and renal tissues. Biochem Biophys Res Commun 180:1298–1303
61. Gasic S, Wagner OF, Vierhapper H, Nowotny P, Waldhausl W (1992) Regional hemodynamic effects and clearance of endothelin-1 in humans: renal and peripheral tissues may contribute to the overall disposal of the peptide. J Cardiovasc Pharmacol 19:176–180
62. Dupuis J, Stewart DJ, Cernacek P, Gosselin G (1996) Human pulmonary circulation is an important site for both clearance and production of endothelin-1. Circulation 94:1578–1584
63. Angg<a>rd E, Galton S, Rae G (1989) The fate of radioiodinated endothelin-1 and endothelin-3 in the rat. J Cardiovasc Pharmacol 13 (Suppl 5):S46-S49
64. Gandhi CR, Harvey SAK, Olson MS (1993) Hepatic effects of endothelin: metabolism of [125I]endothelin-1 by liver-derived cells. Arch Biochem Biophys 305:38–46
65. Löffler BM, Kalina B, Kunze H (1991) Partial characterization and subcellular distribution patterns of endothelin-1, -2 and -3 binding sites in human liver. Biochem Biophys Res Commun 181:840–845
66. Löffler BM, Breu V, Clozel M (1993) Effect of endothelin receptor antagonists and of the novel non-peptide antagonist Ro 46–2005 on endothelin levels in rat plasma. FEBS Lett 333:108–110
67. Fukuroda T, Fujikawa T, Ozaki S, Ishikawa K, Yano M, Nishikibe M (1994) Clearance of circulating endothelin-1 by ETB receptors in rats. Biochem Biophys Res Commun 199:1461–1465
68. Gray GA, Löffler BM, Clozel M (1994) Characterization of endothelin receptors mediating contraction of the rabbit saphenous vein. Am J Physiol 266:H959-H966
69. Teerlink JR, Breu V, Clozel M, Clozel JP (1994) Potent vasoconstriction mediated by endothelin ETB receptors in canine coronary arteries. Circulation 74:105–114
70. Jackman HL, Morris PW, Deddish PA, Skidgel RA, Erdos EG (1992) Inactivation of endothelin I by deamidase (lysosomal protective protein). J Biol Chem 267:2872–2875
71. Jackman HL, Morris PW, Rabito SF, Johansson GB, Skidgel RA, Erdos EG (1993) Inactivation of endothelin-1 by an enzyme of the vascular endothelial cells. Hypertension 21:II 925-II 928
72. Deng AY, Martin LL, Balwierczak JL, Jeng AY (1994) Purification and characterization of an endothelin degradation enzyme from rat kidney. J Biochem 115:120–125
73. Janas J, Sitkiewicz D, Warnawin K, Janas RM (1994) Characterization of a novel, high-molecular weight, acidic, endothelin-1 inactivating metalloendopeptidase from the rat kidney. J Hypertension 22:1155–1162
74. Llorens-Cortes C, Huang H, Vicart P, Gasc JM, Paulin D, Corvol P (1992) Identification and characterisation of neutral endopeptidase in endothelial cells from venous or arterial origins. J Biol Chem 267:14012–14018
75. Perez-Vicario F, Cooper A, Corder R, Fournier A, Warner T (1995) Rapid degradation of endothelin-1 by an enzyme released by the rat isolated perfused mesentery. Br J Pharmacol 114:867–871
76. Patrignani P, Del-Maschio A, Bazzoni G, Daffonchi L (1992) Inactivation of endothelin by polymorphonuclear leukocyte-derived lytic enzymes. Blood 78:2715–2720

77. Maggi CA, Giuliani S, Patacchini R, Rovero P, Giachetti A, Meli A (1989) The activity of peptides of the endothelin family in various mammalian smooth muscle preparations. Eur J Pharmacol 174:23–31
78. Warner TD, De Nucci G, Vane JR (1989) Rat endothelin is a vasodilator in the isolated perfused mesentery of the rat. Eur J Pharmacol 159:325–326
79. Arai H, Hori S, Aramori I, Ohkubo H, Nakamichi S (1990) Cloning and expression of a cDNA encoding an endothelin receptor. Nature 348:730–732
80. Sakurai T, Yanagisawa M, Takuwa Y, Miyazaki H, Kimura S, Goto K, Masaki T (1990) Cloning of a cDNA encoding a non-isopeptide selective subtype of the endothelin receptor. Nature 348:732–735
81. Bax WA Saxena PR (1994) The current endothelin receptor classification: time for reconsideration? Trends Pharmacol Sci 15:379–386
82. Sakamoto A, Yanagisawa M, Sakurai T, Takuwa Y, Yanigasawa H, Masaki T (1991) Cloning and functional expression of human cDNA for the ETB endothelin receptor. Biochem Biophys Res Commun 178:656–663
83. Karne S, Jayawickreme CK, Lerner MR (1993) Cloning and characterization of an endothelin-3 specific receptor (ETC receptor) from Xenopus laevis dermal melanophores. J Biol Chem 268:(19126–19133
84. Kumar C, Mwangi V, Nuthalaguni P, Wu H-L, Pullen M, Brun K, Aiyar H, Morris RA, Naughton R, Nambi P (1994) Cloning and characterization of a novel endothelin receptor from Xenopus heart. J Biol Chem 269:13414–13420
85. Hosada K, Nakao K, Tamura N, Arai H, Ogawa Y, Suga S, Nakanishi S, Imura H (1992) Organization, structure, chromosomal assignment and expression of the human endothelin-A receptor. J Biol Chem 267:18797–18804
86. Arai H, Nakao K, Takaya K, Hosoda K, Ogawa Y, Nakanishi S. Imura H (1993) The human endothelin B receptor gene: structural organization and chromosomal assignment. J Biol Chem 268:3463–3470
87. Frank HJL, Levin ER, Hu RM, Pedram A (1993) Insulin stimulates endothelin binding and action on cultured vascular smooth muscle cells. Endocrinology 133:1092–1097
88. Kanno K, Hirata Y, Tsujino M, Imai M, Shichiri M, Ito H, Marumo F (1993) Upregulation of ETB receptor subtype mRNA by angiotensin II in rat cardiomyocytes. Biochem Biophys Res Commun 194:1282–1287
89. Elshourbagy NA, Korman DR, Wu HL, Sylvester DR, Lee JA, Nuthalaganti P, Bergsma DJ, Kumar CS, Nambi P (1993) Molecular characterization and regulation of the human endothelin receptors. J Biol Chem 268:3873–3879
90. Roubert P, Gillard-Roubert V, Pourmarin L, Cornet S, Guilmard C, Plas P, Pirotzky E, Chabrier PE, Braquet P (1994) Endothelin receptor subtypes A and B are up-regulated in an experimental model of acute renal failure. Mol Pharmacol 45:182–188
91. Winkles JA, Alberts GF, Brogi E, Libby P (1993) Endothelin-1 and endothelin receptor mRNA expression in normal and atherosclerotic human arteries. Biochem Biophys Res Commun 191:1081–1088
92. Yorikane R, Miyauchi T, Sakai S, Sakurai T, Yamaguchi I, Sugushita Y, Goto K (1993)Altered expression of ETB-receptor mRNA in the lung of rats with pulmonary hypertension. J Cardiovasc Pharmacol 22 (Suppl 8):S336–S338
93. Sakurai T, Morimoto H, Kasuya Y, Takuwa Y, Nakauchi H, Masaki T, Goto K (1992) Level of ETB receptor mRNA is down-regulated by endothelins through decreasing the intracellular stability of mRNA molecules. Biochem Biophys Res Commun 186:342–347
94. Shyamala V, Moulthrop THM, Stratton-Thomas J, Tekamp-Olson P (1994) Two distinct human endothelin B receptors generated by alternative splicing from a single gene. Cell Mol Biol Res 40:285–296
95. Ogawa Y, Nakao K, Arai H, Nagakawa O, Hosada K, Suga S-I, Nakanishi S, Imura H (1991) Molecular cloning of a non-isopeptide-selective human endothelin receptor. Biochem Biophys Res Commun 178:248–255
96. Elshourbagy NA, Lee JA, Korman DR, Nuthalaganti P, Sylvester DR, Dilella AG, Sutiphong JA, Bergsma DJ, Kumar CS, Nambi P (1992) Molecular cloning and characterization of the major endothelin receptor subtype in porcine cerebellum. Mol Pharmacol 41:465–473
97. Hashido K, Gamou T, Adachi M, Tabuchi H, Watanabe T, Furuichi Y, Miyamoto C (1991) Truncation of N-terminal extracellular or C-terminal intracellular domains of human ETA receptor abrogated the binding activity to ET-1. Biochem Biophys Res Commun 187:1241–1248

98. Krystek SR, Patel PS, Rose PM, Fisher SM, Kienzle BK, Lach DA, Liu ECK, Lynch JS, Novotny J, Webb ML (1994) Mutation of peptide binding site in transmembrane region of a G-protein-coupled receptor accounts for endothelin receptor subtype selectivity. J Biol Chem 269:12383–12386

99. Adachi M, Furuichi Y, Miyamoto C (1994) Identification of a ligand binding site of the human endothelin-A receptor and specific regions required for ligand selectivity. Eur J Biochem 220:37–43

100. Zhu G, Wu LH, Mauzy C, Egloff AM, Mirzadegan T, Chung FZ (1992) Replacement of lysine-181 by aspartic acid in the third transmembrane region of endothelin type B receptor reduces its affinity to endothelin peptides and sarafotoxin 6c without affecting G protein coupling. J Cell Biochem 50:159–164

101. Sakamoto A, Yanagisawa M, Sawamura T, Enoki T, Ohtani T, Sakurai T, Nakao K, Toyo-oka T, Masaki T (1993) Distinct subdomains of human endothelin receptors determine their selectivity to endothelin A-selective antagonist and endothelin B- selective agonists. J Biol Chem 268:8547–8553

102. Adachi M, Hashido K, Trzeciak A, Watanabe T, Furuichi Y, Miyamoto C (1993) Functional domains of human endothelin receptor. J Cardiovasc Pharmacol 22:S121-S124

103. Hashido K, Adachi M, Gamou T, Watanabe T, Furuichi Y, Miyamoto C (1993) Identification of specific intracellular domains of the human ETA receptor required for ligand binding and signal transduction. Cell Mol Biol Res 39:3–12

104. Kimura S, Kasuya Y, Sawamura T, Shinmi O, Sugita Y, Yanagisawa M, Goto K, Masaki T (1988) Structure-activity relationships of endothelin: importance of the C-terminal moiety. Biochem Biophys Res Commun 156:1182–1186

105. Ihara M, Saeki T, Fukuroda T, Kimura S, Patel AC, Yano M (1992) A novel radioligand [125I]BQ-3020 selective for endothelin ETB receptors. Life Sci 51:47–52

106. Takai M, Unemura I, Yamasaki K, Watakabe T, Fujitani Y, Oda K, Urade Y, Inui T, Yamamura T, Okada T (1992) A potent and specific agonist, Suc-[Glu9,Ala11,15]-endothelin-1(8–21), IRL 1620, for the ETB receptor. Biochem Biophys Res Commun 184:953–959

107. Williams DL, Jones KL, Pettibone DJ, Lis EV, Clineschmidt BV (1991) Sarafotoxin S6c: an agonist which distinguishes between endothelin receptor subtypes. Biochem Biophys Res Commun 175:556–561

108. Ihara M, Fukuroda T, Saeki T, Nishikibe M, Kojiri K, Suda H, Yano M (1991) An endothelin receptor ETA antagonist isolated from Streptomyces Misakiensis. Biochem Biophys Res Commun 178:132–137

109. Spinella MJ, Malik AB, Everitt J, Anderson TT (1991) Design and synthesis of a specific endothelin 1 antagonist: effects on pulmonary vasoconstriction. Proc Natl Acad Sci USA 88:7443–7446

110. Ihara M, Noguchi K, Saeki T, Fukuroda T, Tsuchida S, Kimura S, Fukami T, Ishikawa K, Nishikibe M, Yano M (1992) Biological profiles of highly potent novel endothelin antagonists selective for the ETA receptor. Life Sci 50:247–255

111. Moreland S (1994) BQ-123, a selective endothelin ETA receptor antagonist. Cardiovasc Drug Rev 12:48–69

112. Ishikawa K, Ihara M, Noguchi K, Mase T, Mino N, Saeki T, Fukuroda T, Fukami T, Ozaki S, Nagase T, Nishikibe M, Yano M (1994) Biochemical and pharmacological profile of a potent and selective endothelin B-receptor antagonist, BQ-788. Proc Natl Acad Sci USA 91:4892–4896

113. Cody WL, Doherty AM, He JX, De PPL, Rapundalo ST, Hingorani GA, Major TC, Panek RL, Dudley DT, Haleen SJ, La Douceur D, Hill KE, Flynn MA, Reynolds EE (1992) Design of a functional hexapeptide antagonist of endothelin (1). J Med Chem 35:3301–3303

114. Lam KT, Williams DL, Sigmund JM, Sanchez M, Genilloud O, Kong YL, Steven-Miles S, Huang L, Garrity GM (1992) Cochinmicins, novel and potent cyclodepsipeptide endothelin antagonists from a Microbiospora sp. 1. Production, isolation and characterization. J Antibiotics 45:1709–1716

115. Kikuchi T, Ohtaki T, Kawata A, Imada T, Asami T, Masuda Y, Sugo T, Kusomoto K, Kubo K, Watanabe T, Wakimusu M, Fujino M (1994) Cyclic hexapeptide endothelin receptor antagonists highly potent for both receptor subtypes ETA and ETB. Biochem Biophys Res Commun 200:1708–1712

116. Ikeda S, Awane Y, Kusumoto K, Wakimasu M, Watanabe T, Fujino M (1994) A new endothelin receptor antagonist, TAK-044, shows long-lasting inhibition of both ETA- and ETB-mediated blood pressure responses in rats. J Pharmacol Exp Ther 270:728–733

117. Fujimoto M, Mihara S, Nakajima S, Ueda M, Nakamura M, Sakurai K. A novel non-peptide endothelin antagonist isolated from bayberry, Myrica cerifera. FEBS Lett (1992) 305:41–44

118. Ohashi H, Akiyama H, Nishikori K, Mochizuki JI (1992) Asterric acid, a new endothelin binding inhibitor. J Antibiotics, 45, 1684–1685.
119. Clozel M, Breu V, Burri K, Cassal J, Fischli W, Gray GA, Hirth G, Löffler BM, Müller M, Neidhart W, Ramuz H (1993) Pathophysiological role of endothelin as revealed by the first orally active endothelin receptor antagonist. Nature 365:759–761
120. Ohlstein EH, Nambi P, Douglas SA, Edwards RM, Gellai M, Lago A, Leber JD, Cousins RD, Gao A, Frazee JS, Peishoff CE, Bean JW, Eggleston DS, Elshourbagy NA, Kumar C, Lee JA, Yue TL, Louden C, Elliott JD (1994) SB 209670, a rationally designed potent nonpeptide endothelin receptor antagonist. Proc Natl Acad Sci USA 91:8052–8056
121. Stein PD, Hunt JT, Floyd DM, Moreland S, Dickinson K, Mitchell C, Liu E, Webb ML, Murugesan N, Dickey J, McMullen D, Zhang R, Lee VG, Serafino R, Delaney C, Schaeffer TR, Kozlowski M (1994) The discovery of sulfonamide endothelin antagonists and the development of the orally active ETA antagonist 5-(dimethylamino) -N-(3,4-dimethyl-5- isoxazolyl)-1-naphthalenesulfonamide. J Med Chem 37:329–331
122. Hosada K, Nakao K, Arai H, Suga S-I, Ogawa Y, Mukoyama M, Shirakami G, Saito Y, Nakanishi S, Imura H (1991) Cloning and expression of human endothelin-1 receptor cDNA. FEBS Lett 287:23–26
123. Cristol JP, Warner TD, Thiemermann C, Vane JR (1993) Mediation via different receptors of the vasoconstrictor effects of endothelins and sarafotoxins in the systemic circulation and renal vasculature of the anaesthetized rat. Br J Pharmacol 108:776–779
124. Moreland S, McMullen DM, Delaney CL, Lee VG, Hunt JT (1992) Venous smooth muscle contains vasoconstrictor ETB-like receptors. Biochem Biophys Res Commun 184:100–106
125. Sumner MJ, Cannon TR, Mundin JW, White DG, Watts IS (1992) Endothelin ETA and ETB receptors mediate vascular smooth muscle contraction. Br J Pharmacol 107:858–860
126. Clozel M, Gray GA, Breu V, Löffler BM, Osterwalder R (1992) The endothelin ETB receptor mediates both vasodilation and vasoconstriction in vivo. Biochem Biophys Res Commun 186:867–873
127. Davenport AP, O'Reilly G, Molenaar P, Maguire J, Kuc R, Sharkey A, Bacon C, Ferro A (1993) Human endothelin receptors characterized using reverse transcriptase-polymerase chain reaction, in situ hybridization, and subtype-selective ligands BQ123 and BQ3020: evidence for expression of ETB receptors in human vascular smooth muscle. J Cardiovasc Pharmacol 22:S22–S25
128. Davenport AP, Maguire JJ (1994) Is endothelin-induced vasoconstriction mediated only by ETA receptors in humans? Trends Pharmacol Sci 15:9–11
129. Moreland S, Abboa-Offei B, Seymour A (1994) Evidence for a differential location of vasoconstrictor endothelin receptors in the vasculature. Br J Pharmacol 112:704–708
130. Tschudi MR, Lüscher TF (1994) Characterization of contractile endothelin and angiotensin receptors in human resistance arteries: evidence for two endothelin and one angiotensin receptor. Biochem Biophys Res Commun 204:685–690
131. Deng L-Y, Li J-S, Schiffrin EL (1995) Endothelin receptor subtypes in resistance arteries from humans and rats. Cardiovasc Res 29:532–535
132. Eguchi S, Hirata Y, Imai T, Kanno K, Marumo F (1994) Phenotypic change of endothelin receptor subtype in cultured rat vascular smooth muscle cells. Endocrinology 134:222–228
133. Batra VK, McNeill JR, Xu Y, Wilson TW, Gopalakrishnan V (1993) ETB receptors on aortic smooth muscle cells of spontaneously hypertensive rats. Am J Physiol 264:C479-C484
134. Clozel M, Breu V, Gray GA, Kalina B, Löffler BM, Burri K, Cassal JM, Hirth G, Müller M, Neidhart W, Ramuz H (1994) Pharmacological characterization of bosentan, a new orally active non-peptide endothelin receptor antagonist. J Pharmacol Exp Ther 270:228–235
135. Warner TD, Allcock GH, Vane JR (1994) Reversal of established responses to endothelin-1 in vivo and in vitro by the endothelin receptor antagonists, BQ-123 and PD 145065. Br J Pharmacol 112:207–213
136. Bodelsson G, Sternquist M (1993) Characterization of endothelin receptors and localization of [125I]-endothelin-1 binding sites in human umbilical artery. Eur J Pharmacol 249:299–305
137. Stanimirovic DB, McCarron RM, Spatz M (1994) Dexamethasone down-regulates endothelin receptors in human cerebromicrovascular endothelial cells. Neuropeptides 26:145–152
138. Frelin C, Ladoux A, Marsault R, Vigne P (1991) Functional properties of high- and low-affinity receptor subtypes for ET-3. J Cardiovasc Pharmacol 17 (Suppl 7):S131-S133
139. Molenaar P, O'Reilly G, Sharkey A, Kuc RE, Harding DP, Plumpton C, Gresham GA, Davenport AP (1993) Characterization and localization of endothelin receptor subtypes in the human atrioventricular conducting system and myocardium. Circulation 72: 526–538

140.Schomisch-Moravec C, Reynolds EE, Stewart RW, Bond M (1989) Endothelin is a positive inotropic agent in human and rat heart in vitro. Biochem Biophys Res Commun 159:14–18
141. Jones LG, Rozich JD, Tsutsui H, Cooper G (1992) Endothelin stimulates multiple responses in isolated adult ventricular cardiac myocytes. Am J Physiol 263:H1447-H1454

Endothelin Antagonists:
Novel Treatments for Hypertension?

G. A. GRAY, E. J. MICKLEY and P. E. McEWAN

A role for endothelin-1 (ET-1) in the pathogenesis of hypertension has been postulated since its first description as a potent vasoconstrictor agent derived from the vascular endothelium (Yanagisawa et al. 1988). When administered intravenously in experimental animals or man ET-1 produces long-lasting increases in blood pressure (Yanagisawa et al. 1988; Clarke et al. 1989; Vierhapper et al. 1990). ET-1 can also modulate blood pressure by potentiation of responses to other vasoconstrictor agents and enhancement of sympathetic activity, as well as by stimulation of the generation of renin, angiotensin II, aldosterone and adrenaline (for review see Ferro and Webb 1996). All of these factors and the finding that ET-1 is a co-mitogen with potential to be a mediator of the vascular and cardiac hypertrophy that is characterisitic of hypertension (Schiffrin 1995) lend powerful support to the argument that ET-1 is an important contributor to the pathogenesis of hypertension. Early attempts to use plasma concentrations of ET-1 as an index of the role of ET-1 in experimental and clinical hypertension produced varying, but mostly negative results (see discussion below). Several early studies also demonstrated altered sensitivity to ET-1 in blood vessels from models of hypertension, but it was not until the first antagonists of ET receptors were described that real proof was provided that ET receptors might be a valid therapeutic target in hypertension. The aim of this chapter is to briefly review the experimental evidence for a role of ET-1 in hypertension and to consider the possibility that ET antagonists might provide a novel treatment for clinical hypertension. For further discussion of the role of ET-1 in hypertension the reader is referred to several review articles including Vanhoutte 1993; Luscher et al 1993; Schiffrin 1995; Ferro and Webb 1996 and Gray and Webb 1996.

Endothelin Formation in Hypertension

Investigation of the circulating concentration of immunoreactive ET-1 in animal models of hypertension has yielded varied results (Gray and Webb 1996). When compared to normotensive rats, ET-1 levels were found to be similar in deoxycorticosterone acetate (DOCA)-salt sensitive rats (Suzuki et al. 1990; Lariviere et al. 1993a) or decreased in spontaneously hypertensive rats (SHR; Suzuki et al. 1990; Onho et al. 1992) and transgenic [(mRen-2)27] hypertensive rats (Gardiner et al. 1995). Increased plasma ET-1 levels have only been convincingly demonstrated in models of malignant hypertension (Nishikabe et al. 1993) or where renal dysfunction is present (Vemulapalli et al.

1991; Kohno et al. 1991). These results are reflected in clinical studies. Well characterised hypertensive patients with normal renal function have similar circulating ET-1 concentrations as normotensive patients (Davenport et al. 1990; Schiffrin and Thibault 1991; Haak et al. 1992), but patients with impaired renal and pulmonary function have increased levels (Cernacek and Stewart 1989; Kohno et al. 1990).

The relevance of plasma levels of ET-1 is called into question by the fact that ET-1 is secreted abluminally by endothelial cells to its site of action on underlying vascular smooth muscle cells (Wagner et al. 1992). ET-1 found in the plasma does not therefore provide an accurate representation of functionally active ET-1 but rather represents overflow which has not been cleared from the circulation. Studies of local ET-1 production in tissues by measurement of ET-1 mRNA expression or ET-1 immunoreactivity are more likely to give an indication of the role of ET-1 in animal models of hypertension. Increased ET-1 formation has been observed in aorta and mesenteric arteries of DOCA-salt sensitive (Lariviere et al. 1993a; Fujita et al. 1995; Day et al. 1995b) and DOCA-salt treated SHR (Schiffrin et al. 1995b). In the DOCA-salt sensitive rat there is a significant correlation between aortic immunoreactive ET-1 and systolic blood pressure (Fujita et al. 1995), although plasma levels of ET-1 are not increased in this model (Suzuki et al. 1990; Lariviere et al. 1993b). Furthermore, infusion of phosphoramidon, an inhibitor of ET-1 generation by endothelin-converting enzyme (ECE), in DOCA rats results in a fall in blood pressure greater than that seen in the normotensive control (Vemulapalli et al. 1993). In the Goldblatt one-kidney, one clip (1K, 1C) non-renin dependent model of hypertension in the rat, expression of ET-1 mRNA coincides with the appearance of vascular hypertrophy in the aorta, mesenteric arteries and heart (Sventek et al. 1996). The relationship between tissue level of ET-1 and hypertension or hypertrophy does not however hold true for all models of hypertension. In the SHR model, ET-1 formation was found to be increased in renal tissue (Kitamura et al. 1989; Hughes et al. 1992) but not in the aorta and mesenteric bed (Lariviere et al. 1993a) despite increased blood pressure relative to the normotensive Wistar-Kyoto (WKY) rat. Northern blot analysis confirmed that there were no differences in ET-1mRNA turnover between SHR and WKY aortas (Lariviere et al. 1995a), suggesting that ET-1 formation is likely to be regulated independently in the renal tissue and aorta. The importance of increased ET-1 generation in the SHR is also called into question by the fact that the hypotension following infusion of the ECE inhibitor phosphoramidon was similar in SHR and WKY (McMahon et al. 1991). Vascular and cardiac hypertrophy are also present without evidence for increased ET-1 expression in the renin-dependent 2 K, 1C model of hypertension (Sventek et al. 1996). This highlights the fact that evidence for a role of ET-1 in experimental hypertension is highly dependent on the model used.

Responses to Endothelin

Lack of evidence for increased ET-1 production does not rule out a role for ET-1 as a mediator of hypertension. Even if local ET formation is not increased ET receptor antagonists would still be of use if increased sensitivity to ET-1 was a contributory factor to increased blood pressure. This could occur as a result of increased ET receptor expression or of expression of an alternative type of receptor with higher sensitivity. It is important therefore to characterise not only responsiveness to ET-1 but also the receptors mediating responses to ET-1. Responses to ET-1 and its isopeptides ET-2 and ET-3 are mediated by two major subtypes of receptor, the ET-1 selective ET_A receptor and the isopeptide unselective ET_B receptor. Both types of receptor are found on vascular smooth muscle and can mediate constriction and although the ET_A receptor is the normally the major type found physiologically, there is evidence that the ET_B receptor, which is generally more sensitive, is found in increased amounts pathophysiologically (for review see Gray and Webb 1996). The ET_B receptor is also found on the vascular endothelium where it mediates the release of prostacylin and nitric oxide, vasodilators which counteract the constrictor effect of ET-1 on vascular smooth muscle. Many studies of receptor sensitivity have been carried out using tissue from experimental models of hypertension and as with the studies mentioned above the results are highly variable and highly dependent on the tissue and model studied, as well as the investigator. At least part of this variability might arise because of the fact that contractile responses tend to be amplified in vessels that have undergone vascular hypertrophy (Folkow 1978), a confounding factor that cannot be ignored in any study of vascular responsiveness in blood vessels from hypertensive animals.

In vivo, systemic doses of ET-1 are reported to have enhanced pressor effects in SHR and DOCA-salt sensitive models of hypertension, in comparison to normotensive animals (Miyanchi et al. 1989; Roberts-Thomson et al. 1994). This correlates with reports of increased sensitivity to ET-1 in several tissues from these models, including aorta (Clozel 1989; Suzuki et al. 1990), mesenteric vessels (De Carvalho et al. 1990), renal arteries (Tomobe et al. 1988), portal veins (Kamata et al. 1990) and the coronary vasculature (Takeshita et al. 1991). It should be noted however that a number of investigators also report reduced sensitivity to ET-1 in aortic rings (Auch-Schwelk and Vanhoutte 1989; Criscione et al. 1990), isolated mesenteric arteries (Dohi and Luscher 1991) and renal arteries (Auch-Schwelk and Vanhoutte 1989) from SHR and mesenteric arteries from 2K, 1C rats (Dohi et al. 1991). In human subcutaneous resistance arteries from mild essential hypertensive patients the maximum constriction to ET-1 is significantly reduced, a phenomenon attributed to decreased receptor density, although this was not specifically investigated (Schiffrin et al. 1992). The number of ET-1 binding sites in the membrane fraction of SHR aortas is variously reported to be increased (Tomobe et al. 1991) or decreased (Gu et al. 1990) compared to normotensive WKY rats. The latter would correspond with observations that maximum responses to ET-1 in isolated vessels from hypertensive animals are blunted, even although the sensitivity to ET-1 is increased (Clozel et al. 1989; Nguyen et al. 1992; Deng and Schiffrin 1992). A reduction in the total number of receptors is not inconsistent with increased sensitivity if the relative number of high affinity receptors is increased or if coupling to intracellular signalling mechanisms is enhanced, or both. There have been a number of reports of enhanced elevation of intracellular calcium by ET-1 in smooth muscle cells from hypertensive animals (Gopalakrishnan et al. 1992; Batra et al. 1992). Batra et al. (1992) attributed the increased sensitivity of calcium responses to increased expression of the higher affinity ET_B type receptors for ET-1 by the smooth muscle cells or to more efficient coupling of the ET_B receptor to its effectors. Seo and

Luscher (1995) have also reported increased constriction to the ET_B receptor selective agonist, sarafotoxin S6c in renal arteries from SHR compared to WKY rats and Kuwuhara et al. (1996) found an increased pressor response to the ET_B receptor agonist IRL 1720 in anaesthetised SHR. A possible mechanism for upregulation of ET_B receptors in hypertension is suggested by Barber et al. (1996). They found that fistula induced chronic increases in blood flow in canine femoral arteries resulted in the appearance of a constrictor response to sarafotoxin S6c. Receptor binding studies and immuno-histochemical staining against ET_B receptors confirmed upregulation of ET_B receptors. ET_B receptors are also reported to be upregulated in atherosclerosis, a condition frequently associated with high blood pressure (Dagassan et al. 1995). Upregulation of ET_B receptors is not a generalised feature in all vascular beds. In mesenteric arteries from SHR, both constriction and the evoked $[Ca^{2+}]_i$ response to sarafotoxin S6c are reported to be reduced compared to normotensive controls (Touyz et al. 1995a). Cultured mesenteric arterial smooth muscle cells from SHRs also showed reduced ET_B receptor mediated $[Ca^{2+}]_i$ responses and this was attributed to decreased receptor density (Touyz et al. 1995b).

Patients with essential hypertension exhibit enhanced venoconstriction to ET-1 and a potentiation of sympathetically mediated venoconstriction by infused ET-1 that is not present in normotensive subjects (Haynes et al. 1994). The receptors mediating this effect have not thus far been characterised.

Effects of ET Receptor Blockade on Blood Pressure in Hypertension

As discussed above, the effects of ET-1 are mediated by the ET_A and ET_B subtypes of receptors. There are now a range of peptide and non-peptide, orally active, selective and non-selective antagonists of these receptors (Table 1). A number of these antagonists have been used as tools to investigate the potential role of ET-1 in hypertension.

Table 1. Selected endothelin receptor antagonists

Receptor selectivity		Antagonist	Reference
ET_A	Peptide	BQ-123	Ihara et al. 1991
		TTA-386	Kitada et al. 1993
	Non-peptide	BMS 182874	Stein et al. 1994
		PD155080	Doherty et al 1995
		A127722	Winn et al. 1996
ET_B	Peptide	BQ-788	Ishikawa et al. 1994
	Non-peptide	Ro 46–8443	Breu et al. 1996
ET_A/ET_B	Peptide	PD145065	Cody et al. 1993
		TAK-044	Kikuchi et al. 1994
	Non-peptide	Bosentan	Clozel et al. 1994
		SB209670	Ohlstein et al. 1994
		L754,142	Williams et al. 1995

The ET_A receptor antagonist, BQ-123 was the first compound to be used extensively in experimental animals (Moreland 1994). Continuous intravenous infusion of BQ-123 in vivo in SHR results in a drop in blood pressure (Nishikibe et al. 1993; Ohlstein et al. 1993; McMahon et al. 1993). However, in at least one of these studies (Nishikibe et al. 1993), a similar reduction in blood pressure is seen in normotensive WKY, suggesting that the effect of BQ-123 represents a hypotensive rather than an anti-hypertensive effect. Other studies suggest that BQ-123 and other ET_A receptor antagonists, including BMS-18274 (Bird et al. 1995) can have anti-hypertensive effects, particularly in non-renin dependent models of hypertension, like the DOCA-salt treated SHR (Okada et al. 1994). BQ-123 was reported to have no effect on blood pressure in either the 2K, 1C or aortic banding models of hypertension, both of which are characterised by high levels of renin (Bazil et al. 1992). The most consistently positive results have been achieved in the DOCA-salt sensitive rats with ET_A receptor blockade (Bird et al. 1995; Fujita et al. 1995), non-selective ET_A/ET_B receptor blockade (Li et al. 1994) and ECE inhibition (Vemulapalli et al. 1993) all resulting in large decreases in blood pressure. The non-peptide ET_A/ET_B receptor antagonist, bosentan produces a large reduction of blood pressure in DOCA salt SHR (Schiffrin et al. 1995b) but has no significant effect on the blood pressure of control SHR (Li and Schiffrin 1995).

Ro 46–8443 is a novel non-peptide antagonist with selectivity for the ET_B receptor (Breu et al. 1996) that has been used to investigate the role of ET_B receptors in three different models of hypertension; SHR, DOCA-salt sensitive and rats made hypertensive by chronic administration of L-nitroarginine methyl ester (L-NAME), an inhibitor of the synthesis of the endothelium-derived vasodilator nitric oxide (NO, Clozel and Breu 1996). In normotensive rats Ro 46–8443 was found to decrease blood pressure, consistent with a role for constrictor ET_B receptors in mediating a vasoconstrictor effect of ET-1 (Fig. 1). However, in SHR and DOCA rats the antagonist caused an increase in blood pressure (Fig. 1). This is most likely caused by inhibition of the endothelial ET_B receptor mediated release of relaxing factors, including NO, because in L-NAME treated rats, Ro 46–8443 also produced a reduction in blood pressure. This study therefore appears to reveal upregulation of vasorelaxant endothelial ET_B receptors in SHR and DOCA-salt sensitive rats. In normotensive and L-NAME-hypertensive rats the prevailing role of ET_B receptors is to mediate vasoconstrictor tone. The results are not necessarily inconsistent with reports of upregulated smooth muscle ET_B receptors in experimental hypertension. The role of the constrictor ET_B receptors in the regulation of blood pressure in SHR and DOCA rats may be hidden by the dominance of the endothelial ET_B receptors and perhaps also because of cross-talk between the smooth muscle ET_A and ET_B receptors. This phenomenon, which is well documented in a number of tissues (Clozel and Gray 1995; Fukuroda et al. 1997; Mickley et al., 1997), allows ET_A receptors to compensate upon blockade of ET_B receptors or vice versa (Fig. 2). Under these conditions effective blockade of the response to ET-1 is only achieved when both receptors are blocked simultaneously. This might also explain why bosentan, which blocks all ET receptors, has a greater hypotensive effect in SHR than the ET_A receptor antagonist, BQ-123. There is recent evidence that ET receptor antagonists can cause vasodilatation in hypertensive patients (Ferro et al. 1996) and bosentan has been reported to lower the blood pressure of patients with essential hypertension (Schmitt et al. 1995).

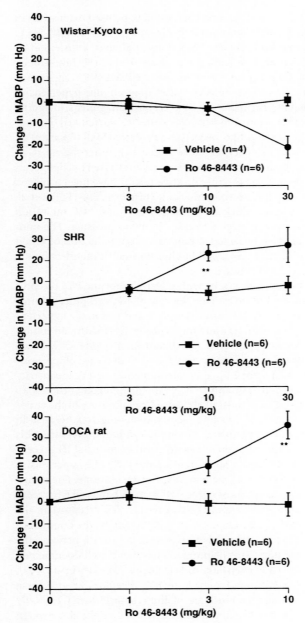

Fig. 1. The effect of intravenous administration of the ET_B receptor antagonist Ro 46–8443 or its vehicle on mean arterial blood pressure *(MABP)* in anaesthetised Wistar-Kyoto rats, SHR and DOCA hypertensive rats. *$p < 0.05$, **$p < 0.01$ compared with vehicle. (From Clozel and Breu 1996)

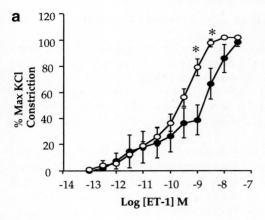

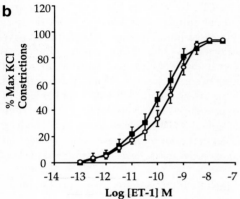

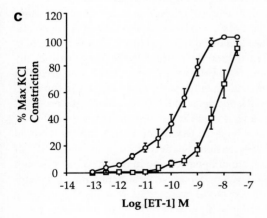

Fig. 2a–c. The effect of **a** ET_A receptor blockade with BQ-123 (*filled circles; n = 8*), **b** ET_B receptor blockade, using BQ-788 (*filled squares; n = 8*) or c blockade of ET_A and ET_B receptors, using BQ-123 and BQ-788 (*open squares; n = 8*) on constrictor responses to ET-1 in rat small mesenteric arteries. *$p < 0.05$ compared to control (*open circles*) ET-1 responses. Blockade of both receptor subtypes is required to effectively inhibit the response to ET-1. (Modified from Mickley et al., 1997)

Endothelin and End-Organ Damage in Hypertension

ET-1, in addition to its role as a systemic vasoconstrictor peptide, is a growth factor within the cardiovascular system (Battistini et al. 1993; Schiffrin 1995). A number of studies in vitro have shown that ET-1 induces vascular smooth muscle cell (VSMC) mitogenesis (Hirata et al. 1989), hypertrophy (Bobik et al. 1990) and protein synthesis (Chua et al. 1992). The known long-term growth-promoting effects of ET-1 support the hypothesis that ET-1 might be a mediator of end-organ damage in hypertension, which includes vascular and cardiac hypertrophy.

Several experimental models of hypertension have been used to investigate the role of ET-1 as a growth factor in vivo. In the DOCA salt rat model of hypertension, increased ET-1 mRNA expression and immunoreactivity (Lariviere et al. 1993a) are associated with increased vascular hypertrophy (Schiffrin et al. 1996). The main controversy to arise from these findings is whether the elevated ET-1 levels are due to hypertension-induced cell damage or to upregulation of ET-1 by a physiological mechanism and, secondly, whether vascular hypertrophy in this animal model is a direct growth effect of ET-1 or a secondary consequence of increased blood pressure. Many studies have attempted to distinguish the direct humoral actions of ET-1 from changes which arise due to shear stress or pressure (Lariviere et al. 1993b; Karam et al. 1996a,b). Increased ET-1 expression in the DOCA-salt rat may occur because of hypertension-induced endothelial cell damage; however, mesenteric arteries from SHR with a similar level of hypertension showed either no difference or reduced expression of ET-1 compared to the normotensive WKY controls (Lariviere et al. 1993b). Furthermore, in SHR given salt or DOCA-salt, ET-1 expression increased in the aorta and mesenteric arteries and exhibited a degree of hypertrophy that was excessive for the level of systolic blood pressure (Schiffrin et al. 1995).

Long-term administration of the combined ET_A/ET_B receptor antagonist, bosentan, to DOCA-salt SHR, whilst only slightly lowering blood pressure, blunted the excessive vascular hypertrophy observed in small mesenteric arteries (Schiffrin et al. 1995), suggesting specific growth-promoting effects of ET-1 in cases of mineralocorticoid excess. In the DOCA-salt rat, however, bosentan had no significant effect on medial thickness in the aorta (Karam et al. 1996a). In a further study in which the SHR was chronically treated with the NO synthase inhibitor L-NAME to increase blood pressure, ET-1 mRNA expression was enhanced in the endothelium of the aorta but not in the small mesenteric arteries (Li et al. 1996). The lack of change in ET-1 expression also correlated with absence of severe hypertrophy, again linking ET-1 to the growth process. Although ET-1 expression correlates with vascular hypertrophy in the DOCA salt rat it would appear that not all animal models of hypertension are affected similarly. The Goldblatt 2K, 1C rat develops a renin-dependent form of hypertension in which ET-1 expression is increased and vascular hypertrophy develops. However, treatment with ET antagonists does not regress hypertrophy or lower blood pressure (Li et al. 1996) suggesting that in this particular animal model the growth effects of ET-1 are probably not as important as those of angiotensin II.

A recent study on the normotensive Sprague-Dawley rat showed that the ET system may be involved in the pathogenesis of neointimal formation (Wang et al. 1996). The authors demonstrated increased ECE-1, prepro ET-1, preproET-3, ET_A and ET_B receptor mRNA expression in carotid arteries following balloon angioplasty and the subsequent formation of the neointima, thus implicating the ET system in this pathological growth process. The role of the ETs in this process is confirmed by the observation that treatment of rats with the dual ET_A/ET_B receptor antagonist SB209670 prevents the formation of the neointima (Douglas et al. 1995).

If ET-1 does have a role as a mediator of hypertrophy it is likely to be a relatively complex one. ET-1 is not only a co-mitogen that can regulate the formation and action of other growth factors (Battistini et al. 1993), but its own synthesis and actions are controlled by other systemic peptides and growth factors. The majority of studies have focussed on the growth interactions between the ET system and the renin-angiotensin system. In vitro, Ang II induces the expression of ET_B (but not ET_A) receptor mRNA in cultured cardiac myocytes (Kanno et al. 1993). Ang II also increases ET-1 expression and ET-1 secretion in cultured vascular smooth muscle cells (Hahn et al. 1990). Furthermore, Ito et al. (1993) have shown that ET-1 can also upregulate Ang II receptors, demonstrating a high degree of synergism between the two systems. In vivo, Ang II increases prepro endothelin mRNA in the mesenteric resistance arteries of the SHR and, via the production of endothelin, indirectly augments contractile responses to norepinephrine in an endothelium-dependent manner (Dohi et al. 1992). This may also involve long term changes in vascular structure. There is also emerging evidence to suggest a role for steroids in the regulation of the ET system. The synthetic glucocorticoid agonist RU28362 has been shown to reduce the number of ET_A receptors on cultured VSMC (Provencher et al. 1995). Although glucocorticoids cause loss of weight gain in rats and induce apoptosis in many cell types (Wyllie et al. 1980) the mechanisms are not known and is possible that growth inhibition results from antagonising the growth effects of systemic growth factors such as ET-1. These hypotheses merit further investigation.

Human and experimental hypertension are associated with complex cardiac remodelling combining cardiac hypertrophy (Frohlich et al 1983), a decrease of coronary vascular reserve (Wicker et al. 1983; Yamamoto et al. 1985) and an increase in interstitial and perivascular fibrosis (Pearlman et al. 1982; Weber et al. 1988; Brilla et al. 1990). In vitro, the actions of ET-1 on cardiac growth are convincing. ET-1 induces hypertrophy of cardiomyocytes assciated with the induction of muscle-specific gene transcripts (Ito et al. 1991). Cultured rat cardiomyocytes express prepro-ET-1 transcripts and release mature ET-1 into the culture medium (Ito et al. 1993). Endothelial cells modulate both cardiac fibroblast collagen synthesis and degradation via ET-1 (Guarda et al. 1993). In vivo, the actions of ET-1 are less clear, however more evidence is emerging to suggest a role for ET-1 in cardiac hypertrophy and remodelling. BQ 123, an ET_A selective receptor antagonist has been shown to block the increase in left ventricular weight due to haemodynamic overload (Ito et al. 1994). Cardiac hypertrophy is a prominent feature in the DOCA salt hypertensive rat model and is associated with an enhancement or over-expression of the ET-1 gene and an increase in the amount of immunoreactive ET-1 in the heart (Lariviere et al. 1995). However, enhanced cardiac ET-1 expression occurs mostly in the endothelium of large and small coronary arteries and in the myocardium. Therefore, it is suggested that enhanced ET-1 expression could participate in coronary microvascular hypertrophy rather than in myocardial hypertrophy (Schiffrin et al. 1995).

Karam et al. (1996b) recently suggested a more definitive role for ET-1 in cardiac remodelling. Also using the DOCA salt rat model, they showed that the ET_A/ET_B antagonist, bosentan, markedly reduced cardiac hypertrophy, perivascular fibrosis and subendocardial fibrosis at a dose which had little effect on reducing arterial pressure (Fig. 3). The reduction in left ventricular hypertrophy with bosentan strongly suggests

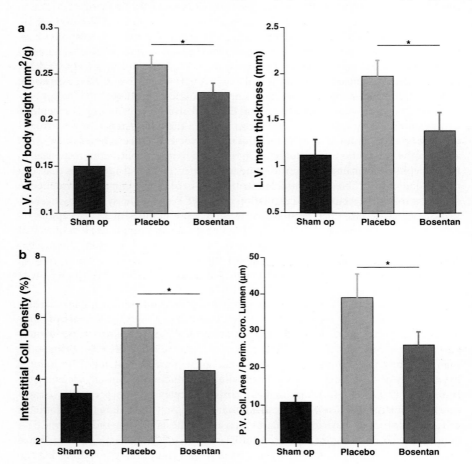

Fig. 3a,b. Inhibition of **a** left ventricular hypertrophy and **b** left ventricular fibrosis in DOCA-salt hypertension by a non-hypotensive dose of the ET_A/ET_B receptor antagonist, bosentan (from the study of Karam et al. 1996b). Three groups of rats were used, sham operated *(Sham-op)*, DOCA treated (one kidney removed) with placebo in food admix *(Placebo)*, and DOCA treated with an oral dose of bosentan as food admix *(Bosentan)*. At the end of the treatment period arterial blood pressure was significantly raised in the placebo treated DOCA group (240 ± 6 mmHg) and the bosentan treated DOCA group (225 ± 5 mmHg) compared to the sham operated group (155 ± 2 mmHg). *$p < 0.05$ compared to placebo group. (Figure kindly supplied by Dr Volker Breu, F. Hoffmann-La Roche)

the ET receptor antagonists have the potential to decrease cardiac hypertrophy by a mechanism which appears to act independently of blood pressure (Karam et al. 1996 a, c). This hypothesis is reinforced by the recent finding that long term treatment with the ET_A receptor antagonist BQ123 prevented left ventricular remodeling that occurred following coronary artery ligation in the rat (Sakai et al. 1996).

The precise mechanisms of action of ET-1 in the heart remain unclear. It remains unknown as to whether ET-1 promotes cardiomyocyte proliferation in vivo. Indeed, there is some controversy as to whether myocytes proliferate or merely increase in size. Cardiac fibroblasts, which are phenotypically different from terminally differentiated myocytes, do proliferate in vitro in response to ET-1 (Kanno et al 1993) but their contribution to left ventricular hypertrophy in vivo is, as yet, unknown. The local interactions of ET-1 with the renin-angiotensin system are also unclear. However, there is evidence to suggest that Ang II inhibits the proliferation of myoendocardial cells which may be a potential source of ET-1 within the heart (Stoll et al. 1995).

Although evidence in man is limited, in pathological states such as coronary transplantation, there is a strong correlation between ET-1 mRNA expression and immunoreactivity within sites of regeneration and granulation and ET-1 levels have been shown to be higher in areas of neovascularisation and angiogenesis (Giaid et al. 1995). ET-1 may not be the only ET isopeptide with growth promoting effects; molecular genetics studies on the Dahl salt sensitive hypertensive rat model have provided some evidence that the ET-3 gene but not the ET-1 gene locus is genetically linked on chromosome 3 and cosegregates with blood pressure and heart weight (Cicila et al. 1994).

The Therapeutic Potential of Inhibitors of the Endothelin System in Hypertension

The experimental models of hypertension discussed in this chapter can be split into two broad groups, those which are characterised by high renin activity, which include the aortic ligation and the Goldblatt 2 kidney, 1 clip models and those which have hypertension with low renin, including the genetically hypertensive SHR and the mineralocorticoid DOCA-salt model. In general, evidence for involvement of ET-1 in the pathophysiology of hypertension is greatest in the low renin models and in particular the DOCA salt rat. In this model, blood pressure is reduced by ET receptor antagonists or inhibitors of ET synthesis; local ET-1 synthesis appears to be increased in some tissues and regions of increased ET-1 synthesis co-localise with regions of vascular or cardiac hypertrophy. Several studies have shown that hypertrophy can occur independently of increased blood pressure, suggesting that induction of local growth factors is responsible. One of these factors is angiotensin II, a peptide produced in large amounts in high renin hypertension. Thus, even if acute reduction of blood pressure by ET receptor antagonists is limited, chronic treatment might have a role in prevention of the hypertrophy that leads to prolonged increases in blood pressure and increased risk of associated complications in renin as well as non-renin dependent hypertension. Only acute studies have been carried out to date in clinical hypertension, but show positive results (Schmitt et al. 1995; Ferro et al. 1996). Bosentan, a non-peptide, ET_A/ET_B receptor antagonist acutely lowers blood pressure in patients with essential hypertension (Schmitt et al. 1995). Although this might simply be a hypotensive effect, as ET receptor antagonists also reduce blood pressure in normotensives due to inhibition of basal release of ET-1 by vascular endothelial cells (reviewed by Ferro and Webb 1996), it does demonstrate the potential utility of ET receptor antagonists in hypertension. Future studies will show if long-term treatment with ET receptor antagonists has advantages over the current anti-hypertensive therapies. Several of the potent antagonists currently in development are ET_A receptor selective antagonists and future studies will also show whether these are as effective as mixed receptor antagonists, bearing in mind the experimental evidence for upregulation of ET_B type receptors in hypertension.

Acknowledgements. The authors thank the British Heart Foundation, the Medical Research Council and the Wellcome Trust for support of their research. Volker Breu, Martine Clozel and Louise Sutherland are acknowledged for their help in compilation of this manuscript.

References

Auch-Schwelk W, Vanhoutte, PM (1989) Contractions to endothelin in isolated arteries of spontaneously hypertensive rats and Wistar Kyoto rats. FASEB 3 A1008

Barber DA, Michener SR, Ziesmer SC, Miller VM (1996) Chronic increases in blood flow upregulate endothelin-B receptors in arterial smooth muscle. Am J Physiol 270 H65–H71

Batra VK, McNeil, JR Xu Y Wilson TW, Gopalakrishnan V (1993) Endothelin-B receptors on aortic smooth muscle cells of spontaneously hypertensive rats. Am J Physiol 264 (33) C479–C484

Battistini B, Chailler P, D'Orleans-Juste P, Briere N, Sirois P (1993) Growth regulatory properties of endothelins. Peptides 14:385–399 USA

Bazil K, Lappe RW, Webb RL (1992) Pharmacological characterization of an endothelinA receptor antagonist in conscious rats J Cardiovasc Pharmacol 20:940–948

Bird JE, Moreland S, Waldron, TL, Powell JR (1995) Antihypertensive effects of a novel endothelin-A receptor antagonist in rats. Hypertension 25:1191–1195

Bobik A, Grooms A, Millar JA, Mitchell A, Grinpukel S (1990) Growth factor activity of endothelin on vascular smooth muscle. Am J Physiol 258:C408–C415

Breu V, Clozel M, Burri K, Hirth G, Niedhart W, Ramuz H (1996) In vitro characterisation of Ro 46–8443, the first non-peptide antagonist selective for the endothelin ETB receptor. FEBS Letters 383:37–41

Brilla CG, Pick R, Tan LB, Janicki JS, Weber KT (1990) Remodelling of the rat right and left ventricles in experimental hypertension. Circ Res 67:1355–1364

Cernacek P, Stewart DJ (1989) Immunoreactive endothelin in human plasma: marked elevations in patients in cardiogenic shock. Biochem Biophys Res Commun 161:562–567

Chua BHL, Krebs CJ, Chua CC, Diglio CA (1992) Endothelin stimulates protein synthesis in smooth muscle cells. Am J Physiol 262:E412–E416

Cicila GT, Rapp JP, Bloch KD, Kurtz TW, Pravenec M, Kren V, Hong CC, Quertermous T, Ng SC (1994) Co-segregation of the endothelin-3 locus with blood pressure and relative heart weight in inbred Dahl rats. J. Hypertens. 12:643–651

Clarke JG, Benjamin N, Larkin SW, Webb DJ, Keogh BE, Davies GJ, Maseri A (1989a) Endothelin is a potent longlasting vasoconstrictor in man. Am J Physiol 257:H2003–H2035

Clozel M (1989) Endothelin sensitivity and receptor binding in the aorta of spontaneously hypertensive rats. J Hypertens 7:913–917

Clozel M, Breu V (1996) The role of receptors ETB in normotensive and hypertensive rats as revealed by the non-selective ETB receptor antagonist Ro 46–8443. FEBS Letts 383:42–45

Clozel M, Gray GA (1995) Are there different ETB receptors mediating constriction and relaxation? J Cardiovasc Pharmacol 26 (Suppl 3):S262–S264

Clozel M, Breu V, Gray GA, Kalina B, Loffer BM, Burri K, Cassal JM, Hirth G, Muller M, Niedhart W, Ramuz H (1994) Pharmacological characterization of bosentan, a new orally active non peptide endothelin receptor antagonist. J Pharmacol Exp Ther 270:228–235

Cody WL, Doherty AM, He JX, De Pue PL, Waite LA, Topliss JG, Haleen SJ, La Douceur D, Flynn MA, Hill KE, Reynolds EE (1993) The rational design of a highly potent combined ETA and ETB receptor antagonist (PD145065) and related analogues. Med Chem Res 3:154–1622

Criscione L, Nellis P, Riniker B, Thomann H, Burdet, R (1990) Reactivity and sensitivity of mesenteric vascular beds and aortic rings of spontaneously hypertensive rats to endothelin: effects of calcium entry blockers Br J Pharmacol 99:31–36

Dagassan PH, Breu V, Clozel M, Kunzli A, Vogt P, Turina M, Kiowski W, Clozel JP (1996) Up-regulation of endothelin B receptors in atherosclerotic human coronary arteries. J Cardiovasc Pharmacol 27:147–153

Davenport AP, Ashby MJ, Easton P, Ella S, Bedford J, Dickerson C, Nunez DJ, Capper SJ, Brown MJ (1990) A sensitive radioimmunoassay measuring endothelin-like immunoreactivity in human plasma: comparison of levels in patients with essentail hypertension and normotensive control subjects. Clin Sci 78:261–264

Day R, Lariviere R, Schiffrin EL (1995a) In situ hybridization shows increased endothelin-1 m RNA levels in endothelin cells iof blood vessels of deoxycorticosterone acetate salt hypertensive rats. Am J Hypertens 19:753–757

Day R, Lariviere R, Schiffrin EL (1995b) In situ hybridisation shows increased endothelin-1 levels in endothelial cells of blood vessels of DOCA-salt hypertensive rats. Am. J. Hypertension 8:294–300

De Carvalho MHC, Nigro D, Scivolette R, Barbeiro V, de Oliveira MA, de Nucci, G, Fortes ZB (1990) Comparison of the effect of endothelin on microvessels and macrovessels in Goldblatt II and DOCA-salt hypertensive rats. Hypertension Suppl 24. I68–I71

Deng LY, Schiffrin EL (1992) Effects of endothelin on resistance arteries of DOCA-salt hypertensive rats. Am J Physiol 262:H1782–H1787

Doherty AM, Patt WC, Repine J, Edmunds JJ, Berryman KA, Reisdorf BR, Walker D M, Haleen SJ, Keiser JA, Flynn MA, Welch KM, Hallak H, Reynolds EE (1995) Structure-activity relationships of a novel series of orally non peptide ETA and ETA/B endothelin receptor-selective antagonist. J Cardiovasc Pharmacol 26 (Suppl 3):S358–S361

Dohi Y, Luscher TF (1991) Endothelin in hypertensive resistance arteries: intraluminal and extraluminal dysfunction. Hypertension 18:543–549

Dohi Y, Criscione L, Luscher TF (1991) Renovascular hypertension impairs formation of EDRF and sensitivity to endothelin-1 in resistance arteries. Br J Pharmacol 104:349–354

Dohi Y, Hahn AWA, Boulanger CM, Buhler FR, Luscher TF (1992) Endothelin stimulated by angiotensin II augments contractility of spontaneoulsy hypertensive rat resistance arteries. Hypertension 19:131–137

Douglas SA, Vickery Clark LM, Louden C, Elliot JD, Ohlstein EH (1995) Endothelin receptor subtypes in the pathogenesis of angioplasty-induced neointima formation in the rat: A comparison of selective ET (A) receptor antagonism and dual ET (A)/ET (B) receptor antagonism using BQ123 and SB209670 dual (A/B) receptor antagonism in the rat common carotid artery. J Cardiovasc Pharmacol 26:S186–S189

Ferro CJ, Webb DJ (1996) The clinical potential of endothelin receptor antagonists in cardiovascular medecine. Drugs 51:12–27

Ferro CJ, Haynes WG, Hand MF, Webb DJ (1996) Are the vascular endothelin and nitric oxide systems involved in the pathophysiology of essential hypertension? Eur J Clin Invest 26 (suppl.1):A51

Frohlich ED (1983) Hemodynamic and other determinants in development of left ventricular hypertrophy. Fed Proc 42:2709–2715

Fujita K, Matsumura Y, Kita S, Miyazaki Y, Hisaki K, Takaoka M, Morimoto S (1995) Role of endothelin-1 and the ETA receptor in the maintenance of deoxycorticosterone acetate-salt-induced hypertension. Br J Pharmacol 114:925–930

Fukuroda T, Ozaki S, Ihara M, Ishikawa K, Yano M, Miyauchi T, Ishikawa S, Goto K, Nishikibe M (1996) Necessity of dual blockade of endothelin ETA and ETB receptor subtypes for antagonism of endothelin-1- induced contraction in human bronchi. Br J Pharmacol 117:995–998

Gardiner SM, March JE, Kemp PA, Mullins JJ, Bennett T (1995) Haemodynamic effects of losartan and the endothelin antagonist SB 209670 in conscious, transgenic ((mRen-2)27) hypertensive rats. Br J Pharmacol 116:2237–2244

Giaid A, Saleh D, Yanagisawa M, Clarke Forbes RD (1995) Endothelin-1 immunoreactivity and mRNA in the transplanted human heart. Transplantation 59:1308–1313

Gopalakrishnan V, Xu YJ, Sharma VK, McNeill JR, Wilson TW, Shen SS (1992) Endothelin-1 but not vasopressin and angiotensin II evoked [Ca^{2+}]i is higher in aortic smooth muscle cells of spontaneously hypertensive rats. J Hypertension 10: (Suppl 4), S81

Gray GA, Webb DJ (1996) The Endothelin system and its potential as a therapeutic target in cardiovascular disease. Pharmacol 72:109–148

Gu XH, Casley DJ, Cincotta M, Nayler WG (1990) 125I-Endothelin-1 binding to brain and cardiac membranes from normotensive and spontaneously hypertensive rats. Eur J Pharmacol 177:205–209

Guarda E, Myers PR, Brilla CG, Tyagi SC, Weber KT (1993) Endothelial cell induced modulation of cardiac fibroblast collagen metabolism. Cardiovasc Res 27:1004–1008

Haak T, Junmanne, Felber A, Hillman U, Usadel KH (1992) Increased plasma levels of endothelin in diabetic patients with hypertension. Am J Hypertens 5:161–166

Hahn AWA, Resink TJ Scott-Burden T, Powell JS, Dohi Y, Buhler FR (1990) Stimulation of endothelin mRNA and secretion in rat vascular smooth muscle cells: a novel autocrine function. Cell Regul 1:649–659

Haynes WG, Hand M, Johnston H, Padfield PL, Webb DJ (1994) Direct and sympapthetically mediated venoconstriction in essential hypertension: enhanced responses to endothelin-1. J Clin Invest 94:1359–1364

Hirata Y, Tagaki Y, Fukuda Y, Marumo F (1989) Endothelin is a potent mitogen for rat vascular smooth muscle cells. Atherosclerosis 78:225–228

Ihara M, Fukuroda T, Saeki T, Nishikibe M, Kojiri K, Suda H, Yano M (1991) An endothelin receptor ETA antagonist isolated from Streptomyces misakiensis.Biochem Biophys Res Commun 178:132–137

Ishikawa K, Ihara M, Noguchi K, Mase T, Mino N, Saeki M, Yano M (1994) Biochemical and Pharmacological profile of a potent and selective endothelin B-receptor antagonist BQ-788. Proc Natl Acad Sci USA 91:4892–4896

Ito H, Hirata Y, Hiroe M (1991) Endothelin-1 induces hypertrophy with enhanced expression of muscle-specific genes in cultured neonatal rat cadiomyocytes. Clin Res. 69:209–215

Ito H, Hirata Y, Adachi S, Tanaka M, Tsujino M, Koike A, Nogami A, Murumo F, Hiroe M (1993) Endothelin-1 is an autocrine/paracrine factor in the mechanism of angiotensin II-induced hypertrophy in cultured rat cardiomyocytes. J Clin Invest 92:398–403

Ito H, Hiroe M, Hirata Y, Fujisaki H, Adachi S, Akimoto H, Marumo F (1994) Endothelin ETA receptor antagonist blocks cardiac hypertrophy provoked by heamodynamic overload. Circulation 89:2198–2203

Kamata K, Miyata N, Katsuya Y (1990) Effects of endothelin on the portal vein from spontaneously hypertensive and Wistar-Kyoto rats. Gen Pharmacol 21:127–129

Kanno M, Hirata Y, Tsujino M, Imai T, Shirichi M, Ito H, Marumo F (1993) Up-regulation of ET-B receptor subtype mRNA by angiotensin II in rat cardiomyocytes. Bichem Biophys Res Commun 194:1282–1287

Karam H, Heudes D, Gonzales MF, Loffler BM, Clozel M, Clozel JP (1996a) Respective role of humoral factors and blood pressure in aortic remodeling of DOCA hypertensive rats. Am J Hypertens 9:991–998

Karam H, Heudes D, Hess P, Gonzales MF, Loffler BM, Clozel M, Clozel JP (1996b) Respective role of humoral factors and blood pressure in cardiac remodeling of DOCA hypertensive rats. Cardiovasc Res 31:287–295

Karam H, Heudes D, Bruneval P, Gonzales MF, Loffler BM, Clozel M, Clozel JP (1996c) Endothelin antagonism in end-organ damage of spontaneously hypertensive rats. Comparison with angiotenin-converting enzyme inhibition and calcium antagonism. Hypertension 28:379–385

Kikuchi T, Ohtaki T, Kawata A, Imada T, Asami T, Masuda Y, Sugo T, Kusomoto K, Kubo K, Wanatabe T, Wakimusu M, Fujino M (1994) Cyclic hexapeptide endothelin receptor antagonists highly potent for both receptor subtypes ETA and ETB. Biochem Biophys Res Commun 200:1708–1712

Kinoshita O, Kawano Y, Yoshi H, Ashida T, Yoshida K, Akabane S, Kuramochi M, Omae T (1991). Acute and chronic effects of anti-endothelin-1 antibody on blood pressure in spontaneously hypertensive rats. J Cardiovasc. Pharmacol 17 (Suppl. 7):S511–S513

Kitada C, Ohtaki T, Masuda Y, Nomura H, Asami T, Matsumoto Y, Fujino M (1993) Design and synthesis of ETA receptor antagonists and study of ETA receptor distribution. J Cardiovasc Pharmacol 22 (suppl.8):S128-S131

Kitazono T, Heistad DD, Farachi FM (1995) Enhanced responses of the basilar artery to activation of endothelin-B receptors in stroke-prone spontaneously hypertensive rats. Hypertension 25:490–494

Kohno M, Yasumari K, Murakawa K-I, Yokokawa K, Horio T, Fukui T, Takeda T (1990) Plasma immunoreactive endothelin in essential hypertension. Am J Med 88:614–618

Kohno M, Murakawa K-I, Horio T, Yokokawa K, Yasunari K, Fukui T, Takeda T (1991) Plasma immunoreactive endothelin-1 in experimental malignant hypertension. Hypertension 18:93–100

Kuwahara M, Masuda T, Tsubone H, Sugano S, Karaki H (1996) Cardiovascular responses mediated by two types of endothelin ETB receptor in spontaneously hypertensive and wistar- kyoto rats. Eur J Pharmacol 296:55–63

Lariviere R, Day R, Schiffrin EL (1993a) Increased expression of endothelin-1 gene in blood vessels of deoxycorticosterone acetate-salt hypertensive rats. Hypertension 21:916–920

Lariviere R, Thibault G, Schiffrin EL (1993b) Increased endothelin-1 content in blood vessels of deoxycorticosterone acetate-salt hypertensive but not in spontaneously hypertensive rats. Hypertension 21:294–300

Lariviere R, Deng LY, Day R, Sventek P, Thibault G, Schiffrin EL (1995a) Increased endothelin-1 gene expression in the endothelium of coronary arteries and in the endocardium of DOCA-salt hypertensive rats. J Mol Cell Cardiol. 27:2123–2131

Lariviere R, Sventek P, Schiffrin EL (1995b) Expression of endothelin-1 gene in blood vessels of adult spontaneously hypertensive rats. Life Sci 56:1889–1896

Li JS, Schiffrin EL (1995). Effect of chronic treatment of adult spontaneously hypertensive rats with an endothelin receptor antagonist. Hypertension 25:495–500

Li JS, Lariviere R, Schiffrin EL (1994) Effect of a non-selective endothelin antagonist on vascular re-modelling in DOCA-salt hypertensive rats. Evidence for a role of endothelin in vascular hypertrophy. Hypertension 24:183–188

Li JS, Deng LY, Grove K, Deschepper CF, Schiffrin EL (1996a) Comparison of effect of endothelium antagonism and angiotensin-converting enzyme inhibition on blood pressure and vascular structure in spontaneously hypertensive rats treated with n-nitro-l-arginine methyl ester. Hypertension 28:188–195

Li JS, Knafo L, Turgeon A, Garcia R, Schiffrin EL (1996b) Effect of endothelin antagonism on blood pressure and vascular structure in renovascular hypertensive rats. Am J Physiol 271:H88–H93

Lüscher T, Bong-gwan Seo, Buhler FR (1993) Potential role of endothelin in hypertension. Controversy on endothelin in hypertension. Hypertens. 21:752–757

McMahon EG, Palomo MA, Moore WM (1991) Phosphoramidon blocks the pressor activity of big endothelin [1–39] and lowers blood pressure in spontaneously hypertensive rats. J Cardiovasc Pharmacol 17 (Suppl. 7):S29–S33

McMahon EG, Palomo MA, Brown MA, Bertenshaw SR, Carter JS (1993) Effect of phosphoramidon (endothelin converting enzyme inhibitor) and BQ-123 (endothelin receptor subtype A antagonist) on blood pressure in hypertensive rats. Am J Hypertens 6:667–673

Mickley EJ, Gray GA, Webb DJ (1997) Activation of endothelin ETA receptors masks the constrictor role of endothelin ETB receptors in isolated rat small mesenteric arteries. Br J Pharmacol, in press

Miyauchi T, Ishikawa T, Tomobe Y, Yanagisawa M, Kimura S, Sugishita Y, Ito I, Goto K, Masaki, T (1989) Characteristics of the pressor response to endothelin in spontaneously hypertensive and Wistar-Kyoto rats. Hypertension 14:427–434

Moreland S (1994) BQ-123 a selective endothelin ETA receptor antagonist. Cardiovasc Drug Rev 12:48–69

Nguyen PV, Parent A, Deng LY, Fluckiger JP, Schiffrin EL (1992) Endothelin vascular receptors and responses in DOCA-salt hypertensive rats. Hypertension 19 (Suppl II):II98–104

Nguyen PV, Yang X-P, Li G, Deng LY, Fluckiger JP, Schiffrin EL (1993) Contractile responses and signal transduction of endothelin-1 in aorta and mesenteric vasculature of adult spontaneously hypertensive rats. Can J Physiol Pharmacol 71:473–483

Nishikibe M, Tsuchida S, Okada M, Fukuroda T, Shimamoto K, Yano M, Ishikawa K, Ikemoto F (1993) Antihypertensive effect of a newly synthesized endothelin antagonist BQ-123 in a genetic hypertensive model. Life Sci 52:717–724

Ohlstein EH, Douglas SA, Ezekiel M, Gellai M (1993) Antihypertensive effects of the endothelin receptor antagonist BQ-123 in conscious spontaneously hypertensive rats. J Cardiovasc Pharmacol 22 (Suppl 8):S321–S324

Ohlstein EH, Nambi P, Douglas SA, Edwards RM, Gellai M, Lago A, Leber JD, Cousins RD, Gao A, Frazee JS, Peishoff CE, Bean JW, Eggleston DS, Elshourbagy NA, Kumar C, Lee JA, Yue TL, Louden C, Elliot JD (1994) SB 209670 a rationally designed potent non peptide endothelin receptor antagonist. Proc Natl Acad Sci USA 91:8052–8056

Ohno A, Naruse M, Kato S, Hosaka M, Naruse K, Demura H, Sugino N (1992) Endothelin-specific antibodies decrease blood pressure and increase glomerular filtration rate and renal plasma flow in spontaneously hypertensive rats. J Hypertens 10:781–785

Okada M, Fukuroda T, Shimamoto K, Takahashi R, Ikemoto F, Yano M, Nishikibe M (1994) Anti-hypertensive effects of BQ-123 a selective endothelin ETA receptor antagonist in spontaneously hypertensive rats treated with DOCA-salt. Eur J Pharmacol 259:339–342

Pearlman ES, Weber KT, Janicki JS, Pietra GG, Fishman AP (1982) Muscle fiber orientation and connective tissue in the hypertrophied human heart. Lab Invest 46:158–164

Provencher PH, Saltis J, Funder JW (1995) Glucocorticoids but not mineralocorticoids modulate endothelin-1 and angiotensin II binding in SHR vascular smooth muscle cells. J Steroid Biochem Mol Biol 52:219–225

Raschak M, Unger L, Riechers H, Klinge D (1995) Receptor selectivity of endothelin antagonists and prevention of vasoconstriction and endothelin induced sudden death. J Cardiovasc Pharmacol 26 (Suppl 3):S397–S399

Roberts-Thomson, P, McRichie, R, Chalmers, R (1994) Experimental hypertension produces diverse changes in the regional vascular responses to endothelin-1 in the rabbit and the rat. J Hypertens 12:1225–1234

Roubert P, Gillard V, Plas P, Chabrier PE, Braquet P (1990) Down-regulation of endothelin binding sites in rat vascular smooth muscle cells. Am J Hypertension 3:310–312

Sakai S, Miyauchi T, Kobayashi M, Yamaguchi I, Goto K, Sugishita Y (1996) inhibition of myocardial endothelin pathway improves long-term survival in heart failure. Nature 384:353–355

Schiffrin EL (1995) Endothelin: Potential role in hypertension and vascular hypertrophy. Hypertension 25:1135–1143

Schiffrin EL, Thibault G (1991) Plasma endothelin in human essential hypertension. Am J Hypertens 4:303–308

Schiffrin EL, Deng L-Y, Larochelle, P (1992) Blunted effects of endothelin upon subcutaneous resistance arteries of mild essential hypertensive patients. J Hypertens 10 (5):437–444

Schiffrin EL, Lariviere R, Li JS, Sventek P, Touyz RM (1995a) Deoxycorticosterone acetate plus salt induces overexpression of vascular endothelin-1 and severe vascular hypertrophy in spontaneously hypertensive rats. Hypertension 25:769–773

Schiffrin EL, Sventek P, Li Jin-S, Turgeon A, Reudelhuber T (1995b) Antihypertensive effect of an endothelin receptor antagonist in DOCA- salt spontaneously hypertensive rats. Br J Pharmacol 115:1377–1381

Schiffrin EL, Lariviere R, Li JS, Sventek P (1996) Enhanced expression of the endothelin-1 gene in blood vessels of DOCA-salt hypertensive rats: correlation with vascular structure. J Vasc Res. 33:235–248

Schmitt R, Belz GG, Fell D, Lebmeier R, Prager G, Stahnke PL, Sittner WD, Kaworth A, Jones CR (1995) effects of the novel endothelin receptor anatgonist bosentan in hypertensive patients. In: Proceedings of the 7th Meeting of the European Hypertension Society, June, 1995, Milan, Italy

Seo B, Luscher TF (1995) ETA and ETB receptors mediate contraction to endothelin-1 in renal artery of aging SHR: Effects of FR139317 and bosentan. Hypertension 25:501–506

Stein PD, Hunt JT, Floyd DM, Moreland S, Dickinson K, Mitchell C, Liu E, Webb M L, Murugesan N, Dickey J, McMullen D, Zhang R, Lee R, Delaney C, Schaeffer TR, Kozlowski M (1994) The discovery of sulfonamide endothelin antagonists and the development of the orally active ETA antagonist 5-(dimethylamino)-N-(3,4-dimethyl-5-isox-azolyl)-1-napthalenesulfon-amide. J Med Chem 37:329–331

Stoll M, Steckelings M, Paul M, Bottari SP, Metzger R, Unger T (1995) The angiotensin AT2-receptor mediates inhibition of cell proliferation in coronary endothelial cells. J Clin Invest 95:651–657

Suzuki N, Miyauchi T, Tomobe Y, Matsumoto H, Goto K, Masaki T, Fujino M (1990) Plasma concentrations of endothelin-1 in spontaneously hypertensive rats and DOCA-salt hypertensive rats. Biochem Biophys Res Commun 167 (3):941–947

Sventek P, Turgeon A, Garcia R, Schiffrin EL (1996) Vascular and cardiac overexpression of endothelin-1 gene in 1-kidney, 1-clip Goldblatt hypertensive rats but only the late phase of 2-kidney, 1-clip Goldblatt hypertension. J Hypertens 14:57–64

Takeshita H, Nishikibe M, Yano M, Ikemoto F (1991) Coronary vascular response to endothelin in isolated perfused hearts of spontaneously hypertensive rats. Clin Exptal Pharmacol Physiol 18:661–669

Tomobe Y, Miyauchi T, Saito A, Yanagisawa M, Kimura S, Goto K, MasakI T (1988) Effects of endothelin on the renal artery from spontaneously hypertensive and Wistar-Kyoto rats. Eur J Pharmacol 152:373–374

Tomobe Y, Ishikawa T, Yanagisawa M, Kimura S, Masaki T, Goto K (1991) Mechanisms of altered sensitivity to endothelin-1 between aortic smooth muscles of spontaneously hypertensive and Wistar Kyoto rats. J Pharmacol Exp Ther 275:555–561

Touyz RM, Deng L-Y, Schiffrin EL (1995a) Endothelin subtype B-receptor mediated Ca^{2+} and contractile responses in small arteries of hypertensive rats. Hypertension 26:1041–1045

Touyz RM, Lariviere R, Schiffrin EL (1995b) Endothelin receptor subtypes in mesenteric vascular smooth muscle cells of spontaneously hypertensive rats.

Vanhoutte PM (1993) Is endothelin involved in the pathogenesis of hypertension? Hypertens 21: 747–751

Vemulapalli S, Chui PJS, Rivelli M, Foster CJ, Sybertz EJ (1991) Modulation of circulating endothelin levels in hypertension and endotoximia in rats. J Cardiovasc Pharmacol 18:895–903

Vemulapalli S, Watkins RW, Brown A, Cook J, Bernardino V, Chiu PJS (1993) Disparate effects of phosphoramidon on blod pressure in SHR and DOCA-salt hypertensive rats. Life Sci 53:783–793

Vierhapper HO, Wagner P, Nowotny P, Waldhausl W (1990) Effect of endothelin in man. Circulation 81:1415–1418

Wagner OF, Christ G, Wojta J, Vierhapper H, Parzer S, Nowotny PJ, Schneider B, Waldhausl W, Binder BR (1992) Polar secretion of endothelin-1 by cultured endothelial cells. J Biol Chem 267:16066–16068

Wang X, Douglas SA, Louden C, Vickery-Clark LM, Feuerstein GZ, Ohlstein EH (1996) Expression of endothelin-1, endothelin-3, endothelin-converting enzyme-1, and endothelin-A and endothelin-B receptor mRNA after angioplasty-induced neointimal formation in the rat. Circ Res 78:322–328

Weber KT, Janicki JS, Schroff SG, Pick R, Chen RM, Bashey RI (1988) Collagen remodelling of the pressure-overloaded, hypertrophied non-human primate myocardium. Circ Res 62:757–765

Wicker P, Tarazi RC, Kobayashi K (1983) Coronary blood flow during the development and regression of left ventricular hypertrophy in renovascular hypertensive rats. Am J Cardiol 51:1744–1749

Winn M, Von Geldern TW, Opgenorth TJ, Jae HS, Tasker AS, Boyd SA, Kester JA, Mantei RA, Bal R, Sorensen BK, WuWong JR, Chiou WJ, Dixon DB, Novosad EI, Hernandez L, Marsh KC (1996) 2,4-Diarylpyrrolidine-3-carboxylic acids-potent ET(A) selective endothelin receptor antagonists. 1. Discovery of A-127722. J Med Chem 39:1039–1048

Wu CC, Bohr DF (1990) Role of endothelium in the response to endothelin in hypertension. Hypertension 16:677–681

Wyllie AH, Kerr JFR, Currie AR (1980) Cell death: the significance of apoptosis. Int Rev Cytol 68:251–303

Yamamoto J, Tsuchia M, Saito M, Ikeda M (1985) Cardiac contractile and coronary flow reserves in deoxycorticosterone acetate-salt hypertensive rats. Hypertension 7:569–577

Yanagisawa M, Kurihara H, Kimura S, Tomobe Y, Kobayashi M, Mitsui Y, Yazaki Y, Goto K, Masaki T (1988b) A novel potent vasoconstrictor peptiide produced by vascular endothelial cells. Nature 332:411–415

Endothelin in Hypertension: A Role in Vascular Hypertrophy?

E. L. Schiffrin

Abstract

The endothelins are 21-amino acid peptides with potent vasoconstrictor effects. These peptides have as well mild mitogenic and cell hypertrophic effects. Endothelin-1 is the main endothelin produced by the endothelium. Plasma levels of immunoreactive endothelin are normal or only slightly increased in experimental or human hypertension. The responsiveness of blood vessels to endothelin-1 is normal or attenuated in experimental and human hypertension. Endothelin-1 expression is increased in the endothelium of blood vessels in the deoxycorticosterone acetate (DOCA)-salt hypertensive rat and DOCA-salt treated spontaneously hypertensive rats (SHR), in association with exaggerated vascular hypertrophy in comparison to other models of hypertension. Endothelin receptor antagonist treatment attenuates partially the elevation of blood pressure and blunts to an important degree the development of vascular hypertrophy in these experimental models of hypertension, whereas it has no effect on SHR. The latter do not overexpress endothelin in blood vessels. In some severe hypertensive patients there is overexpression of endothelin-1 in endothelium of small arteries. These results suggest that endothelins may play a role in some experimental models of high blood pressure and perhaps in severe human hypertension, in part by accentuating hypertension-induced vascular hypertrophy.

Introduction

Although endothelins were discovered 8 years ago (Yanagisawa et al. 1988), and there has been significant effort to attempt to establish whether these peptides, being powerful vasoconstrictors, could be involved in the pathogenesis of hypertension, it is still unclear whether they participate in the pathophysiology of elevated blood pressure. The development of techniques allowing the demonstration of expression of endothelins in the vascular wall and the discovery of molecules with potent endothelin receptor blocking properties has provided new evidence regarding the potential role of endothelins in hypertension. This data, including some recently available results in human hypertension, will be reviewed.

Endothelin Receptors and Effects on Blood Vessels in Experimental Hypertension

Endothelins interact with ET_A receptors present on smooth muscle cells (Arai et al. 1990) to exert their contractile actions, and ET_B receptors (Sakurai et al. 1990) which are mainly present on the endothelium, where ET_B receptor activation releases nitric oxide and prostacyclin, inducing vascular relaxation. ET_B receptors have been demonstrated to exist as well on smooth muscle cells of aorta (Batra et al. 1993; Panek et al. 1992), pulmonary (MacLean et al. 1994) and coronary arteries (Teerlink et al. 1994) and in veins (Gray et al. 1994; Moreland et al. 1992) of different species. In humans, ET_B receptor-mediated constriction has been shown to occur in mammary arteries (Seo et al. 1994) and in small subcutaneous arteries (Deng et al. 1995). ET_B receptors appear to play a minimal role in small arteries and predominate in veins and pulmonary arteries. However, the contribution of ET_B receptors in resistance arteries could be underestimated, since upon exposure to either endothelin-1 or ET_B selective agonists they rapidly become desensitized (Tschudi and Lüscher 1994; Sharifi and Schiffrin 1996). Endothelin-1 has mild mitogenic and cell hypertrophic properties (Bobik et al. 1990; Hirata et al. 1989), which could contribute to a blood pressure raising effect, by producing vascular hypertrophy.

The responses of the aorta of spontaneously hypertensive rats (SHR) to stimulation by endothelin-1 are reduced (Clozel 1989; Nguyen et al. 1993) or normal (Bolger et al. 1991). In resistance arteries from SHR similar results have been reported (Deng and Schiffrin 1992a; Dohi and Lüscher 1991; Nguyen et al. 1993). When studying deoxycorticosterone acetate (DOCA)-salt hypertensive rats we were intrigued by the profound attenuation of responses to endothelin-1, associated with reduced density of endothelin receptors and intracellular second messenger generation in blood vessels (Flückiger et al. 1992; Nguyen et al. 1992). In resistance arteries dissected from gluteal subcutaneous biopsies from hypertensive humans we also found significantly blunted media stress responses to endothelin-1 (Schiffrin et al. 1992, 1993). Although attenuated, responses to endothelin, as to other agonists, are amplified by the reduced vascular circumference which results from remodeling of small arteries in hypertension. Thus, even if individual smooth muscle cells respond less, vessels may exhibit enhanced responses to endothelins according to the Law of Laplace.

Plasma Levels of Endothelin-1 in Hypertension

Endothelin immunoreactivity has been found to be slightly increased, or often normal, in both experimental and in human hypertension (Davenport et al. 1990; Kohno et al. 1990; Schiffrin and Thibault 1991; Shichiri et al. 1990; Suzuki et al. 1990). Plasma endothelin may be increased in SHR treated with DOCA and salt, and in two-kidney one clip hypertensive rats treated with caffeine, both of which develop malignant hypertension (Kohno et al. 1991). Two cases of hemangioendothelioma overproducing endothelin have been reported in humans (Yokokawa et al. 1991). This is in fact the only example of increased blood-borne endothelin inducing elevated blood pressure in humans.

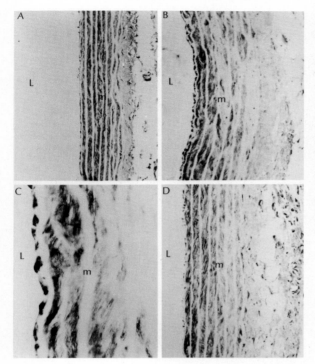

Fig. 1A–D. Photomicrographs show immunohistochemical localization of immunoreactive endothelin-1 in aorta from uninephrectomized (**A**) and deoxycorticosterone acetate-salt hypertensive rats (**B**). A significant increase in immunostaining was detected in the endothelial cell layer of the mesenteric arteries of hypertensive rats (**B, C**). Immunostaining in endothelial cells was almost completely abolished by addition of excess synthetic endothelin-1 (**D**). Magnification **A, B** and **D**, ×250; and **C**, ×1000. *L*, lumen; *m*, media. (From Larivière et al. 1993b)

Vascular Production of Endothelin in Hypertension

Because in most cases blood-borne endothelins did not appear to be involved in hypertension, and since concentrations of endothelin-1 in plasma may not reflect vascular production of endothelin-1, which could act essentially as a paracrine-autocrine agent, we decided to examine the production of endothelins by the vascular wall in hypertension. As mentioned above, we were intrigued by the finding that responses of blood vessels of DOCA-salt hypertensive rats were significantly blunted and the density of endothelin receptors was found to be reduced (Nguyen et al. 1992). We speculated that increases in vascular production of endothelin could induce endothelin receptor downregulation. For this reason immunoreactive endothelin was investigated in blood vessels of DOCA-salt hypertensive rats, which had shown receptor downregulation. We found endothelin to be increased by radioimmunoassay in acid extracts from blood vessels, and by immunohistochemistry in the endothelial layer of the vascular wall in DOCA-salt hypertensive rats (Fig. 1) (Larivière et al. 1993b). Endothelin-1 mRNA measured by Northern blot analysis was also increased in blood vessels of these hypertensive rats (Larivière et al. 1993a). This increase was shown to be of endothelial origin by in situ hybridization in DOCA-salt hypertensive rats (Fig. 2) (Day et al. 1995; Larivière et al. 1995a). In blood vessels of SHR, a similar or lower content of immunoreac-

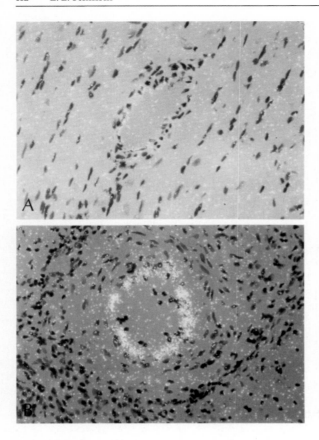

Fig. 2A,B. In situ hybridization of the heart of a control rat (**A**) and of a DOCA-salt hypertensive rat (**B**) with a specific endothelin-1 complementary RNA antisense probe shows heavy labeling of endothelium of a small intramyocardial artery of the latter. No specific enhancement of labeling was detectable in the myocardium. Note the severe hypertrophy of the small intramyocardial artery overexpressing endothelin-1 mRNA, from a DOCA-salt hypertensive rat, in **B.** Original magnification, ×80. Similar results were reported in Larivière et al. (1995a)

tive endothelin was found in SHR in comparison to vessels from normotensive Wistar-Kyoto (WKY) control rats (Larivière et al. 1993b), which agreed with normal or only slightly reduced responses to endothelin and endothelin receptors. We also decided to test whether SHR treated with DOCA and salt would develop enhanced endothelin-1 expression. Indeed, DOCA and salt treatment induced a form of malignant hypertension in SHR, and in association there was enhanced expression of endothelin-1 (Schiffrin et al. 1995d). We also examined expression of endothelin-1 in the vascular wall of renovascular hypertensive rats, and found a moderate elevation of endothelin-1 mRNA in vessels of 1-kidney 1 clip Goldblatt hypertensive rats early on, whereas in 2-kidney 1 clip hypertensive rats elevation could be detected only late in the evolution of hypertension (Sventek et al. 1996b). This is of interest because 1-kidney 1 clip hypertensive rats have low renin activity, whereas 2-kidney 1 clip hypertensive rats initially have high renin activity and later become renin-independent. This suggested, together with the enhanced expression found in DOCA-salt hypertensive rats, that overexpression of endothelin-1 in blood vessels appeared to occur in non renin-dependent experimental rat models. This is particularly interesting considering the apparently paradoxical fact that angiotensin II is a stimulant of endothelin-1 expression by the endothelium (Imai et al. 1992). We have also recently examined the expression of endothelin-1 in L-NAME-induced hypertension, in which no overexpression in blood vessels was found. In the malignant model of hypertension induced by L-NAME in SHR endothelin-1 overexpression was detected only in conduit arteries (Sventek et al. 1996a; Li et al. 1996a).

Endothelin-1 gene overexpression, when present, such as in DOCA-salt hypertensive rats, appears to be a relatively widespread phenomenon, present in aorta and mesenteric arteries as summarized above, but as well in the endothelium of large and small coronary arteries, and areas of endocardium, in the heart (Larivière et al. 1995a), and in the endothelium of renal arteries (Deng et al. 1996). In the kidneys of DOCA-salt hypertensive rats, enhanced expression of endothelin-1 is also found in glomeruli, particularly in mesangial cells.

Relationship of Vascular Endothelin-1 Overexpression and Vascular Hypertrophy

We have also been impressed by the fact that enhanced expression of endothelin-1 was found in blood vessels of DOCA-salt hypertensive rats, in which vascular hypertrophy is very severe (Deng and Schiffrin 1992b; Schiffrin et al. 1996), whereas in SHR, which do not overexpress endothelin-1 in blood vessels, vascular hypertrophy is less severe (Deng and Schiffrin 1992a; Nguyen et al. 1993). This suggested that enhanced expression of endothelin-1 could be the cause for the difference in degree of vascular hypertrophy. If SHR are treated with DOCA and salt to induce malignant hypertension (DOCA-salt SHR), endothelin-1 is overexpressed in blood vessels of SHR, and in this model the vascular hypertrophy found is significantly accentuated (Schiffrin et al. 1995d). In L-NAME-induced hypertension, in which we did not find enhanced endothelin expression, vascular hypertrophy is known to be mild or absent in spite of elevated blood pressure (Dunn and Gardiner 1995; Schiffrin 1995c). In L-NAME-treated SHR, in spite of extremely severe elevation of blood pressure, there is little small artery hypertrophy, but there is some conduit artery hypertrophy. This coincides with overexpression of endothelin-1 in large arteries, but no enhanced expression in small arteries (Sventek et al. 1996a; Li et al. 1996a). Thus, endothelin overexpression and vascular hypertrophy appear to coincide in different models of experimental hypertension (Table 1). When blood pressure elevation and small artery hypertrophy were correlated, it was found that there was a close relationship of media width, media to lumen ratio and media cross-sectional area of small mesenteric arteries in groups of rats with increasing levels of blood pressure elevation except in DOCA-salt hypertensive rats (Fig. 3), in which hypertrophy was exaggerated for the level of blood pressure elevation (Schiffrin et al. 1996). This suggested that endothelin-1 overexpression, present in blood vessels of these rats, could explain the exaggerated vascular hypertrophy in relation to the level of blood pressure. Outer diameter, which is reduced in hypertensive rats ("eutrophic remodeling", see Fig. 3, right upper panel) correlated inversely with blood pressure elevation in all groups, including those in which endothelin-1 expres-

Table 1. Endothelin in different models of hypertension

Hypertensive model	Vascular reactivity	Vascular hypertrophy		Endothelin-1 expression		Response to chronic endothelin receptor antagonism
		Large arteries	Small arteries	Large arteries	Small arteries	
DOCA-salt hypertensive rats	⇓	++++	++++	++++	++++	BP lowering, antihypertrophic
SHR	N	++	++	N	N	None
SHR-sp	?	+++	+++	?	?	BP lowering, antihypertrophic
Malignant DOCA-salt SHR	?	++++	++++	++++	++++	BP lowering, antihypertrophic
2-K 1C Goldblatt	N	+	+	N	N	None
1-K 1C Goldblatt	?	+++	+++	++	?	None
L-NAME	?	+	N	N	N	None
L-NAME-treated SHR	?	+	N	++	N	None
Severe human essential hypertension	⇓	+++	+++	?	++++	?

N, normal.

sion was enhanced, which suggested that whereas hypertrophic remodeling could depend on endothelin-1 overexpression, eutrophic remodeling did not. This has as well been suggested recently in SHR-sp by Chillon et al. (1996) (see below).

Endothelin Antagonists and Experimental Hypertension

In acute experiments, administration of BQ-123, a selective ET_A endothelin receptor antagonist, lowered blood pressure slightly in SHR and in DOCA-salt hypertensive rats (Bazil et al. 1992; McMahon et al. 1993; Nishikibe et al. 1993) but not in renovascular hypertensive rats (Nishikibe et al. 1993). With the development of orally active agents which block both ET_A and ET_B receptors (Clozel et al. 1993), the relationship of vascular hypertrophy, hypertension and endothelin expression has become clearer. Chronic oral treatment of DOCA-salt hypertensive rats with the combined ET_A/ET_B endothelin receptor antagonist bosentan reduced blood pressure by about 20 mmHg, but at the same time blunted the development of vascular hypertrophy of resistance arteries (Fig. 4), beyond what could be explained by the blood pressure lowering (Li et al. 1994). In Fig. 3 it can be noted that bosentan-treated DOCA-salt hypertensive rats exhibited a degree of small artery hypertrophy which correlated with the level of blood pressure

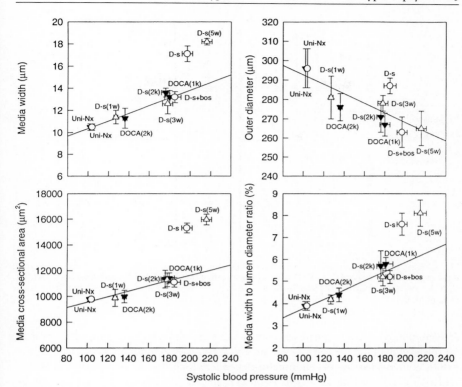

Fig. 3. Correlation of systolic blood pressure and media width, media cross-sectional area, outer diameter and of media width to lumen diameter ratio of mesenteric small arteries. In the case of the media cross-sectional area, although the relationship is theoretically exponential, it was fitted as a linear correlation, since the correlation is a very "flat" exponential one. Groups represented "by closed triangles" include uninephrectomized rats *(Uni-Nx)*, Uni-Nx rats receiving 1% saline to drink (salt), DOCA-treated rats with two kidneys intact *(DOCA-2k)*, DOCA-treated rats which were unilaterally nephrectomized *(DOCA-1k)*, and DOCA-treated rats with two kidneys intact which received 1% saline to drink *(D-s 2k)*. *Open triangles* represent groups overexpressing vascular endothelin-1: DOCA-treated rats, which were uninephrectomized and received 1% saline to drink, that is DOCA-salt hypertensive rats, treated for 1 week *(D-s 1w)*, 3 weeks *(D-s 3w)* or 5 weeks *(D-s)*. Results of DOCA-salt hypertensive rats treated with the combined ET_A/ET_B endothelin receptor antagonist bosentan *(D-s + bos)* or not (D-s). Data from bosentan-treated and -untreated DOCA-salt hypertensive rats and their Uni-Nx controls are depicted in *open circles*. The media width, media cross-sectional area and the media width to lumen diameter ratio of arteries from DOCA-salt hypertensive rats did not correlate with systolic blood pressure, as shown by the method of studentized deleted residuals (since p was >0.05). The media width, media cross-sectional area and media width to lumen diameter ratio of bosentan-treated rats correlated closely with blood pressure. The correlation coefficient for the regression of media width was 0.95, for media cross-section 0.96, and for media to lumen ratio 0.96 after excluding the models overexpressing endothelin-1 ($p < 0.01$). The outer diameter of vessels of all groups correlated with systolic blood pressure, including those models overexpressing endothelin-1 (correlation coefficient=0.64, $p < 0.05$). (From Schiffrin et al. 1996)

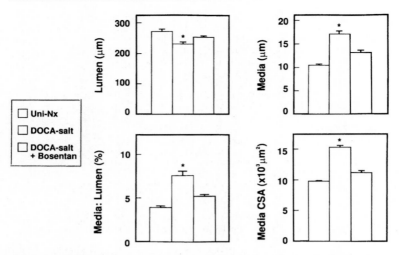

Fig. 4. Standardized parameters measured or calculated in small mesenteric arteries of uninephrectomized rats *(Uni-Nx)*, DOCA-salt hypertensive rats *(DOCA-salt)*, and bosentan-treated DOCA-salt hypertensive rats *(DOCA-salt-Bosentan)*. (From Schiffrin 1995a, re-drawn from data of Li et al. 1994)

elevation, in contrast to untreated DOCA-salt rats, which had excessive hypertrophy for the level of blood pressure elevation. This, as already mentioned, suggested that the enhanced hypertrophy was endothelin-dependent (Schiffrin et al. 1996a). In contrast, "eutrophic" remodeling (reduction of outer diameter of small arteries) appeared to be blood pressure-dependent. Bosentan treatment did not affect cardiac hypertrophy in DOCA-salt hypertensive rats. This may be due to the fact that even though endothelin-1 expression is enhanced in the heart of DOCA-salt hypertensive rats, this is limited to the endothelium of coronary arteries and areas of endocardium (Larivière et al. 1995a). In contrast to DOCA-salt hypertensive rats, chronic treatment with bosentan did not affect blood pressure or vascular hypertrophy in adult SHR (Li and Schiffrin 1995a) nor its development in young SHR (Li and Schiffrin 1995b) which agreed with the absence of vascular overexpression of endothelin-1 in SHR. SHR treated with DOCA and salt, which have enhanced vascular expression of the endothelin-1 gene and severe vascular hypertrophy, responded to chronic endothelin antagonism with blood pressure lowering and reversal of hypertrophy of small arteries (Schiffrin et al. 1995e; Li et al. 1996c). As well, some evidence of a renal protective effect was found in these malignant hypertensive rats after chronic treatment with bosentan, which may relate to the presence of renal vascular and glomerular overexpression of endothelin-1 which was reported in DOCA-salt hypertensive rats (Deng et al. 1996), and is presumably also present in endothelial cells of blood vessels and mesangial cells of glomeruli in the kidney of DOCA-salt SHR. Suggesting that severity of hypertension may be a common denominator for responsiveness to endothelin antagonism, stroke-prone SHR responded to bosentan treatment with lowering of blood pressure and regression of vascular hypertrophy at the level of cerebral arterioles (Chillon et al. 1996). As well, it was observed that "eutrophic" remodeling of cerebral arterioles was unaffected by endothelin antagonism in SHR-sp, in agreement with the studies in DOCA-salt hypertensive rats suggesting that endothelin could be related to hypertrophic remodeling and not to eutrophic remodeling (Schiffrin et al. 1996a). When renovascular hypertension was evaluated with bosentan treatment, neither 1-kidney nor 2-kidney 1 clip Goldblatt hypertensive rats responded to endothelin antagonism (Li et al. 1996b). The result in 1-kidney 1 clip rats was somewhat unexpected, since these rats have moderate overexpression of endothelin-1 in blood vessels (Sventek et al. 1996b). The absence of response may be attributed to the fact that in contrast to DOCA-salt hypertensive rats, overexpression in 1-kidney 1 clip rats is only moderate, and endothelin-dependency may accordingly be a minor component among pressor factors in this hypertensive model. In L-NAME-treated rats, an ET_A selective endothelin antagonist given chronically did not lower blood pressure, as expected from the absence of endothelin overexpression in blood vessels (Sventek et al. 1997). Chronic treatment with bosentan of SHR which have become severely hypertensive after administration of L-NAME did not reduce blood pressure (Li et al. 1996a). It should be noted that these rats exhibited endothelin-1 overexpression only at the level of conduit arteries, and did not have enhanced expression of endothelin-1 or hypertrophy at the level of small arteries. It is possible that overexpression of endothelin-1 in small arteries and endothelin-1-dependent small artery hypertrophy may be necessary conditions for observing a hypotensive response to chronic endothelin receptor antagonist administration (Table 1).

Human Essential Hypertension

There is little data on the role of endothelin in humans with essential hypertension. Most studies have shown that plasma levels of endothelin and vascular responses to endothelin are not enhanced (see above). There are as yet no reports of endothelin receptor antagonist trials in hypertensive patients, but in initial studies there is evidence that systemic endothelin receptor blockade may decrease blood pressure slightly in normotensive humans (Haynes et al. 1996). It is not yet known whether there are increased responses to endothelin receptor antagonists in hypertensive patients. We have studied the expression of endothelin-1 in small arteries of hypertensive patients in comparison to normotensive subjects by *in situ* hybridization in gluteal subcutaneous biopsies. We found that overexpression of endothelin-1 mRNA in the endothelium of small arteries only occurred in the few severe hypertensives investigated, and could not be detected in either normotensive subjects or mild hypertensive patients (Schiffrin et al. 1997). These findings are very preliminary, since very few patients have so far been investigated. To us however, this suggested that as in experimental hypertensive animals, endothelin-1 overexpression occurred in the endothelium of small arteries of individuals with severe hypertension. These patients might respond particularly well to endothelin antagonism, which remains to be demonstrated.

Conclusion

The demonstration of enhanced vascular endothelin-1 expression and the results of treatment with endothelin receptor antagonists indicate a potential involvement of endothelin-1 in the mechanisms of elevated blood pressure and severe vascular hypertrophy in DOCA-salt hypertensive rats, in DOCA-salt treated SHR and in SHR-sp. Preliminary data suggest that overexpression of endothelin-1 may also occur in severe hypertension in humans. Whether this indicates that there are hypertensive patients who might be particularly amenable to treatment with endothelin receptor antagonists remains to be investigated.

Acknowledgments. The work of the author was supported by a Group grant from the Medical Research Council of Canada to the Multidisciplinary Research Group on Hypertension and grants from the Fondation des maladies du coeur du Québec.

References

Arai H, Hori S, Aramori I, Ohkubo H, Nakanishi S (1990) Cloning and expression of a cDNA encoding an endothelin receptor. Nature 348:730–732

Batra KV, McNeill JR, Xu Y, Wilson TW, Gopalkrishnan V (1993) ETB receptors on aortic smooth muscle cells of spontaneously hypertensive rats. Am J Physiol 264:C479–C484

Bazil MK, Lappe RW, Webb RL (1992) Pharmacologic characterization of an endothelinA (ETA) receptor antagonist in conscious rats. J Cardiovasc Pharmacol 20:940–948

Bobik A, Grooms A, Millar JA, Mitchell A, Grinpukel S (1990) Growth factor activity of endothelin on vascular smooth muscle. Am J Physiol (Cell Physiol) 258:C408–C415

Bolger GT, Liard F, Jodoin A, Jaramillo J (1991) Vascular reactivity, tissue levels and binding sites for endothelin: a comparison in the spontaneously hypertensive and Wistar-Kyoto rats. Can J Physiol Pharmacol 69:406–413

Chillon JM, Heistad DD, Baumbach GL (1996) Effects of endothelin receptor inhibition on cerebral arterioles in hypertensive rats. Hypertension 27:794–798

Clozel M (1989) Endothelin sensitivity and receptor binding in the aorta of spontaneously hypertensive rats. J Hypertens 17:913–917

Clozel M, Gray GA, Breu V, Löffler B-M, Osterwalder R (1992) The endothelin ETB receptor mediates both vasodilation and vasoconstriction in vivo. Biochem Biophys Res Commun 186:867–873

Clozel M, Breu V, Burri K, Cassal J-M, Fischli W, Gray GA, Hirth G, Löffler, Müller M, Neidhart W, Ramuz H (1993) Pathophysiological role of endothelin revealed by the first orally active endothelin receptor antagonist. Nature 365:759–761

Davenport AP, Ashby MJ, Easton P, Ella S, Bedford J, Dickerson C, Nunez DJ, Capper SJ, Brown MJ (1990) A sensitive radioimmunoassay measuring endothelin-like immunoreactivity in human plasma: comparison of levels in patients with essential hypertension and normotensive control subjects. Clin Sci 78:261–264

Day R, Larivière R, Schiffrin EL (1995) In situ hybridization shows increased endothelin-1 mRNA levels in endothelial cells of blood vessels of deoxycorticosterone acetate-salt hypertensive rats. Am J Hypertens 8: 294–300

Deng LY, Schiffrin EL (1992a) Effects of endothelin-1 and vasopressin on resistance arteries of spontaneously hypertensive rats. Am J Hypertens 5:817–822

Deng LY, Schiffrin EL (1992b) Effects of endothelin on resistance arteries of DOCA-salt hypertensive rats. Am J Physiol 262:H1782–H1787

Deng LY, Li J-S, Schiffrin EL (1995) Endothelin receptor subtypes in resistance arteries from humans and from rats. Cardiovasc Res 29:532–535

Deng LY, Day R, Schiffrin EL (1996) Localization of sites of enhanced expression of endothelin-1 in the kidney of deoxycorticosterone acetate-salt hypertensive rats. J Am Soc Nephrol (in press)

Dohi Y, Luscher TF (1991) Endothelin in hypertensive resistance arteries: intraluminal and extra-luminal dysfunction. Hypertension 18:543–549

Dunn WR, Gardiner SM (1995) No evidence of vascular remodeling during hypertension indu-ced by chronic inhibition of nitric oxide synthase in Brattleboro rats. J Hypertens 13:849–857

Flückiger JP, Nguyen PV, Li G, Yang XP, Schiffrin EL (1992) Phosphoinositide, calcium and 1,2 diacylglycerol response of blood vessels of DOCA-salt hypertensive rats to endothelin-1. Hypertension 19:743–749

Gray GA, Löffler B-M, Clozel M (1994) Characterization of endothelin receptors mediating con-traction of rabbit saphenous vein. Am J Physiol (Heart Circ Physiol) 266:H959–H966

Haynes WG, Ferro CJ, O'Kane KPJ, Sommerville D, Lomax CC, Webb DJ (1996) Systemic endo-thelin receptor blockade decreases peripheral vascular resistance and blood pressure in humans Circulation 93:1860–1870

Hirata Y, Takagi Y, Fukuda Y, Marumo F (1989) Endothelin is a potent mitogen for rat vascular smooth muscle cells. Atherosclerosis 78:225–228

Imai T, Hirata Y, Emori T, Yanagisawa M, Masaki T, Marumo F (1992) Induction of endothelin-1 gene by angiotensin and vasopressin in endothelial cells. Hypertension 19:753–757

Kohno M, Yasumari K, Murakawa KI, Yokokawa K, Horio T, Fukui T, Takeda T (1990) Plasma immunoreactive endothelin in essential hypertension. Am J Med 88:614–618

Kohno M, Murakawa K-i, Horio T, Yokokawa K, Yasunari K, Fukui T, Takeda T (1991) Plasma immunoreactive endothelin-1 in experimental malignant hypertension. Hypertension 18:93–100

Kurihara H, Yoshizumi M, Sugiyama T, Takaku F, Yanagisawa M, Masaki T, Hamaoki M, Kato H, Yazaki Y (1989) Transforming growth factor-beta stimulates the expression of endothelin mRNA by vascular endothelial cells. Biochem Biophys Res Commun 159:1435–1440

Larivière R, Day R, Schiffrin EL (1993a) Increased expression of endothelin-1 gene in blood vessels of deoxycorticosterone acetate-salt hypertensive rats. Hypertension 21:916–920

Larivière R, Thibault G, Schiffrin EL (1993b) Increased endothelin-1 content in blood vessels of deoxycorticosterone acetate-salt hypertensive but not in spontaneously hypertensive rats. Hypertension 21:294–300

Larivière R, Deng LY, Day R, Sventek P, Thibault G, Schiffrin EL (1995a) Increased endothelin-1 expression in the endothelium of coronary arteries and endocardium in the DOCA-salt hyper-tensive rat J Mol Cell Cardiol 27:2123–2131

Larivière R, Li J-S, Schiffrin EL (1995b) Endothelin-1 expression in blood vessels of DOCA-salt hypertensive rats treated with the combined ETA/ETB endothelin receptor antagonist bosen-tan. Can J Physiol Pharmacol 73: 390–398

Larivière R, Sventek P, Schiffrin EL (1995c) Expression of endothelin-1 gene in blood vessels of adult spontaneously hypertensive rats. Life Sci 56: 1889–1896

Li J-S, Larivière R, Schiffrin EL (1994) Effect of a nonselective endothelin antagonist on vascular remodeling in DOCA-salt hypertensive rats. Evidence for a role of endothelin in vascular hypertrophy. Hypertension 24:183–188

Li J-S, Schiffrin EL (1995a) Effect of chronic treatment with a combined ETA/ETB endothelin receptor antagonist in adult spontaneously hypertensive rats. Hypertension 25 (Part 1):495–500

Li J -S, Schiffrin EL (1995b) Chronic endothelin receptor antagonist treatment of young spontane-ously hypertensive rats. J Hypertens 13:647–652

Li J-S, Deng LY, Grove K, Deschepper CF, Schiffrin EL (1996a) Comparison of effect of endothelin antagonism and angiotensin-converting enzyme inhibition on blood presure and vascular structure in spontaneously hypertensive rats treated with N^{CO}-nitro-L-arginine methyl ester. Correlation with topography of vascular endothelin-1 gene expression. Hypertension 28:188–195

Li J-S, Knafo L, Turgeon A, Garcia R, Schiffrin EL (1996b) Effect of endothelin antagonism on blood pressure and vascular structure in renovascular hypertensive rats. Am J Physiol (Heart Circ Physiol) 40:1188–1193, 19

Li J-S, Schürch W, EL Schiffrin (1996c) Renal and vascular effects of chronic endothelin receptor antagonism in malignant hypertensive rats Amer J Hypertens 9:803–811, 19

MacLean MR, McCulloch KM, Baird M (1994) Endothelin ETA- and ETB-receptor-mediated vaso-constriction in rat pulmonary arteries and arterioles. J Cardiovasc Pharmacol 23:838–845

McMahon EG, Palomo MA, Brown MA, Bertenshaw ST, Carter JS (1993) Effect of phosphorami-don (endothelin converting enzyme inhibitor) and BQ-123 (endothelin receptor subtype A antagonist) on blood pressure in hypertensive rats. Am J Hypertens 6:667–673

Moreland SD, McMullen DM, Delaney CL, Lee VG, Hunt JT (1992) Venous smooth muscle contains vasoconstrictor ETB -like receptors. Biochem Biophys Res Commun 184:100–106

Nguyen PV, Parent A, Deng LY, Flückiger JP, Thibault G, Schiffrin EL (1992) Endothelin vascular receptors and responses in DOCA-salt hypertensive rats. Hypertension 19 (Suppl. II): II–98–II–104

Nguyen PV, Yang X-P, Li G, Deng LY, Flückiger J-P, Schiffrin EL (1993) Contractile responses and signal transduction of endothelin-1 in aorta and mesenteric vasculature of adult spontaneously hypertensive rats. Can J Physiol Pharmacol 71:473–483

Nishikibe M, Tsuchida S, Okada M, Fukuroda T, Shimamoto K, Yano M, Ishikawa K, Ikemoto F (1993) Antihypertensive effect of newly synthesized endothelin antagonist, BQ-123, in a genetic hypertensive model. Life Sci 52:717–724

Okada M, Fukuroda T, Shimamoto K, Takahashi R, Ikemoto F, Yano M, Nishikibe M (1993) Antihypertensive effects of BQ-123, a selective endothelin ETA receptor antagonist, in spontaneously hypertensive rats treated with DOCA-salt. Eur J Pharmacol 259:339–342

Panek RL, Major TC, Hirigorani GP, Doherty AM, Taylor DG (1992) Endothelin and structurally related analogs distinguish between endothelin receptor subtypes. Biochem Biophys Res Commun 183:566–571

Sakurai T, Yanagisawa M, Takuwa Y, Miyazaki H, Kimura S, Goto K, Masaki T (1990) Cloning of a cDNA encoding a non-isopeptide-selective subtype of the endothelin receptor. Nature 348:732–735

Schiffrin EL (1995a) Endothelin, potential role in hypertension and vascular hypertrophy. Hypertension 25: 1135–1143

Schiffrin EL (1995b) Endothelin in hypertension Curr Opinion Cardiol 10:485–489

Schiffrin EL (1995c) Vascular structure in L-NAME-induced hypertension: Methodological considerations for studies of small arteries. J Hypertens 13:817–821

Schiffrin EL, Thibault G (1991) Plasma endothelin in human essential hypertension. Am J Hypertens 4:303–308

Schiffrin EL, Deng LY, Larochelle P (1992) Blunted effects of endothelin on small subcutaneous resistance arteries of mild essential hypertensive patients. J Hypertens 10:437–444

Schiffrin EL, Deng LY, Larochelle P (1993) Morphology of resistance arteries and comparison of effects of vasoconstrictors in mild essential hypertensive patients. Clin Invest Med 16:177–186

Schiffrin EL, Larivière R, Li J-S, Sventek P Touyz RM (1995a) Deoxycorticosterone acetate plus salt induce overexpression of vascular endothelin-1 and severe vascular hypertrophy in spontaneously hypertensive rats. Hypertension 25 [Part 2): 769–773

Schiffrin EL, Sventek P, Li J-S, Turgeon A, Reudelhuber T (1995b) Antihypertensive effect of bosentan, a mixed ETA/ETB endothelin receptor antagonist, in DOCA-salt spontaneously hypertensive rats. Brit J Pharmacol 115:1377–1381

Schiffrin EL, Larivière R, Li J-S, Sventek P (1996) Enhanced expression of endothelin-1 gene in DOCA-salt hypertensive rats: correlation with vascular structure. J Vasc Res 33:235–248

Schiffrin EL, Deng LY, Sventek P, R Day (1997) Enhanced expression of endothelin-1 gene in endothelium of resistance arteries in severe human essential hypertension. J Hypertens 15:57–63

Seo B, Oemar BS, Siebenmann R, Von Segesser L, Lüscher TF (1994) Both ETA and ETB receptors mediate contraction to endothelin-1 in human blood vessels. Circulation 189:1203–1208

Sharifi AM, Schiffrin EL (1996) Endothelin receptors mediating vasoconstriction in rat pressurized small arteries Can J Physiol Pharmacol 74: 934–93

Shichiri M, Hirata Y, Ando K, Emori T, Ohta K, Kimoto S, Ogura M, Inoue A, Marumo F (1990) Plasma endothelin levels in hypertension and chronic renal failure. Hypertension 15:493–496

Suzuki N, Miyauchi T, Tomobe Y, Matsumoto H, Goto K, Masaki T, Fujino M (1990) Plasma concentrations of endothelin-1 in spontaneously hypertensive rats and DOCA-salt hypertensive rats. Biochem Biophys Res Commun 167:941–947

Sventek P, Turgeon A, Schiffrin EL (1997) Vascular endothelin-1 gene expression and effect on blood pressure of chronic ET_A endothelin receptor antagonism after nitric oxide synthase inhibition with L-NAME in normal rats. Circulation 95:240–244

Sventek P, Li J-S, Grove K, Deschepper CF, Schiffrin EL (1996) Vascular structure and expression of endothelin-1 gene in L-NAME-treated spontaneously hypertensive rats. Hypertension 27:49–55

Sventek P, Turgeon A, Garcia R, Schiffrin EL (1996) Vascular and cardiac overexpression of endothelin-1 gene in 1-kidney, one clip Goldblatt hypertensive rats but only in the late phase of 2-kidney, 1 clip Goldblatt hypertension. J Hypertens 14:57–64

Teerlink JR, Breu V, Sprecher U, Clozel M, Clozel J-P (1994) Potent vasoconstriction mediated by endothelin ETB receptors in canine coronary arteries. Circ Res 74:105–114
Tschudi MR, Lüscher TF (1994) Characterization of contractile endothelin and angiotensin receptors in human resistance arteries. Evidence for two endothelin and one angiotensin receptor. Biochem Biophys Res Commun 204:685–690
Yanagisawa M, Kurihara H, Kimura S, Tomobe Y, Kobayashi M, Mitsui Y, Yazaki Y, Goto K, Masaki T (1988) A novel potent vasoconstrictor peptide produced by vascular endothelial cells. Nature 332:411–415
Yokokawa K, Tahara H, Kohno M, Murakawa Ki, Yasunari K, Nakagawa K, Hamada T, Otani S, Yanagisawa M, Takeda T (1991) Hypertension associated with endothelin-secreting malignant hemangioendothelioma. Ann Intern Med 114:213–215

Endothelin in Congestive Heart Failure: Pathophysiology and Therapeutic Implications

D. D. Borgeson, L. J. McKinley and J. C. Burnett, Jr

Summary

Activation of both tissue and circulating endothelin-1 (ET) has emerged as an additional neurohumoral hallmark of congestive heart failure (CHF). The mechanism of activation of ET in CHF probably is multifactorial and includes stimulation by factors such as Angiotensin II and tissue hypoxia. The functional significance of ET activation in CHF is currently being elucidated by the use of potent ET receptor antagonists. Such studies support the concept that ET in CHF contributes to vasoconstriction, alterations in other neurohumoral systems and ventricular hypertrophy. The increase of plasma ET in CHF also has prognostic and diagnostic significance thus supporting its role as a serum marker in the management of CHF. Moreover, the high mortality associated with CHF may be decreased by interrupting the ET system thus supporting the inhibition of ET as a treatment strategy. Thus, ET emerges as both a marker and mediator of CHF as well as a key target in the therapeutics for this important cardiovascular disease.

Introduction

Since 1988, endothelin-1 (ET) has emerged as a unique endothelium-derived cardio-vascular peptide that produces sustained contraction of isolated arteries and veins and proliferation of vascular smooth muscle cells [1–3]. In addition, ET has many diverse biological actions which include increases in myocardial inotropism, regulation of transport of sodium in renal epithelial cells, modulation of other hormonal systems which include atrial natriuretic peptide and renin release as well as activation of myocardial fibroblasts [4–8]. The endothelins which are comprised of three distinct isoforms, ET-1, ET-2, and ET-3 are widely distributed and mediate their biological actions by interacting with at least two receptors which have been cloned and well characterized. The endothelin-A (ET-A) receptor which is widely distributed and preferentially expressed in vascular smooth muscle cells mediates potent vasoconstrictor and growth-promoting actions and binds preferentially to the ET-1 isoform. The endothelin-B (ET-B) receptor is expressed primarily in vascular endothelial cells and binds to all three identified endothelin isoforms and when activated releases nitric oxide and prostacyclin. In addition, ET-B is also present in vascular smooth muscle and may promote vasoconstriction. Thus, ET emerges as an endogneous peptide system which has a functional role as an important cardiovascular regulatory hormone of endothelial origin.

While a role for ET in physiologic regulation continues to emerge, its pathophysiologic importance is suggested by its plasma elevation in disorders of cardiovascular and renal function. Specifically, plasma ET concentrations are increased two to three-fold in congestive heart failure (CHF), two-fold in essential hypertension, three-fold in cardiogenic shock and two-fold in transplantation associated hypertension and radiocontrast-induced nephropathy [9–15]. More recently, its activation at the tissue level has also been reported in experimental CHF, human atherosclerosis and in transplant graft atherosclerosis [16–18].

Studies from our laboratory have focused upon the role of ET in CHF, a syndrome characterized by low cardiac output, sodium and water retention, ventricular dilatation and hypertrophy, and systemic vasoconstriction with activation of the RAAS. Indeed, CHF may serve as a functional paradigm in understanding the role for ET as a local paracrine factor and as a circulating hormone which participates in cardiovascular regulation. Thus, ET complements other vasoconstrictor and mitogenic humoral systems known to be activated in CHF which include the sympathetic nervous system, the RAAS and vasopressin.

Given this activation of ET in heart failure, several questions emerge including the mechanisms of its elevation, the functional significance of its increase, its role as a prognostic marker, its interaction with other vasoactive factors known to be activated in CHF and therapeutic strategies designed to antagonize this cardiovascular peptide. In this review, we will address these questions and speculate as to where future efforts might be directed.

Mechanisms of ET Activation in CHF

A hallmark of overt CHF is the increase in circulating ET which has been demonstrated in both human and experimental CHF [12–15] (Fig. 1). Because ET mRNA is widespread [19], increased ET in CHF could originate from several sources including heart, lung and peripheral vasculature. Indeed, we have reported increased tissue ET in the heart, lung, aorta and kidney in experimental heart failure [16]. Mechanisms for the increase in plasma and tissue ET in CHF could include increased production stimulated by either mechanical or neurohumoral factors, or decreased clearance and/or metabolism of the peptide. With respect to hemodynamic factors, Margulies et al. demonstrated that right and left atrial pressures were independently and significantly correlated with plasma ET in experimental CHF [13]. These correlations are consistent with a report in which the highest level of plasma ET was observed in acute pulmonary edema, presumably in association with marked increases in left atrial pressure [20]. Because recent observations demonstrated pulmonary production of ET in the normal lung, we investigated whether the increase in pulmonary tissue ET in experimental CHF was linked to an increase in ET production as defined by an increase in ET mRNA [21]. In this recent investigation, pulmonary ET was significantly enhanced in a model of CHF.

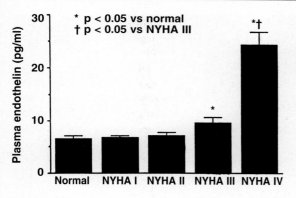

Fig. 1. Plasma endothelin concentration of healthy subjects (*normal*) and congestive heart failure patients. Values are mean ± SEM. *NYHA* indicates New York Heart Association [12]

It is tempting to speculate that elevated cardiac filling pressures, and perhaps chronic cardiac or pulmonary vascular distention, are stimuli for ET production or secretion, as has been demonstrated for atrial natriuretic peptide (ANP) [22]. This concept is also supported by the observation by Cody et al. establishing a strong correlation between circulating ET and pulmonary hypertension in CHF [23]. It should be noted, however, that the increase in pulmonary ET mRNA in our study in experimental CHF was observed in a model of low cardiac output produced by chronic thoracic inferior vena caval constriction (TIVCC). The TIVCC model is unique in that due to the reduction of venous return it is unassociated with atrial distention or increased cardiac filling pressures. Thus an increase in pulmonary pressures are not essential for the elevation of ET in CHF.

A second and equally compelling mechanism for activation of the tissue ET system in CHF is tissue hypoxia. Investigations have established that tissue hypoxia is a potent activator of ET transcription in cultured pulmonary endothelial cells [24]. As tissue hypoxia is a feature of CHF, this may be an additional mechanism for its activation. Indeed, in the intact animal, 45 min of hypoxia which lowered arterial PO_2 to 80% resulted in a marked increase in renal ET both in glomeruli and in renal epithelial cells in the absence of an increase in plasma ET [25] (Fig. 2). Such a role for the kidney as a site of increased production of ET during hypoxia is also suggested by the report by Sandok et al. that during 45 min of decreased renal perfusion pressure produced by suprarenal aortic clamping in the dog results in an increase in plasma ET which is prevented by bilateral nephrectomy [26].

Changes in ET clearance and metabolism might also contribute to increases in circulating ET in CHF. For example, in studies by Cavero and co-workers, low dose exogenous ET produced greater increments in plasma ET in experimental CHF than in normal controls [27]. Moreover, in the TIVCC model, Underwood et al. has demonstrated that whole body metabolic clearance for ET is decreased and may contribute to the striking increases in plasma ET in this setting (unpublished data). In addition, a recent report demonstrated down-regulation of endothelin receptors in cardiac and renal tissue of rabbits with experimental CHF [28]. As a role for the ET-B receptor in the clearance of ET, it has been reported [29] that a down-regulation of "clearance receptors" may be another mechanism contributing to the increases in circulating ET in heart failure.

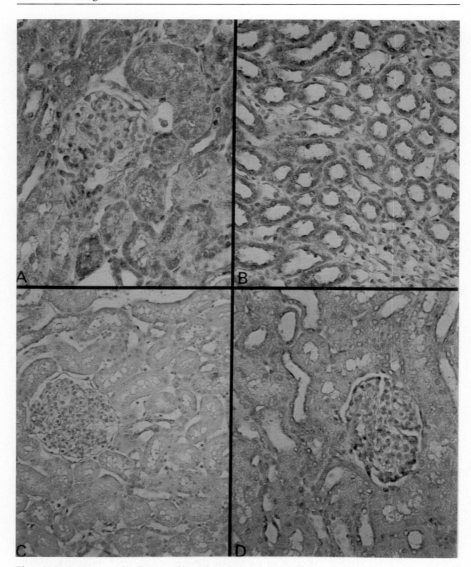

Fig. 2A–D. Representative immunohistochemical staining for ET in kidney of a dog exposed to 1 h of hypoxia or 1 h of room air ventilation. **A** Hypoxic dog cortex. **B** Normoxic dog medulla. **C** Nonimmune control. **D** Normoxic dog cortex. Stained areas represent the presence of tissue immunoreactive ET [25]

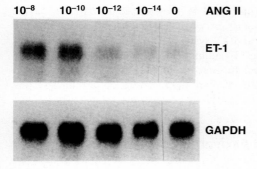

10⁻⁸   10⁻¹⁰   10⁻¹²   10⁻¹⁴   0   ANG II

ET-1

GAPDH

Fig. 3. Dose-response effect of angiotensin II *(Ang II)* on endothelin-1 *(ET-1)* mRNA expression in rat cardiac fibroblasts. Cells were incubated with various doses of Ang II for 30 min. Northern blot analysis was performed in the same manner as in Fig 2 [8]

Repeated investigations have supported a key role for Angiotensin II (ANG II) in the pathophysiology and therapeutics of CHF. Such a role is underscored by clinical trials in which morbidity and mortality are improved by interfering with the generation of ANG II with angiotensin converting inhibitors (ACE-I) [30, 31]. A number of key studies have reported that ANG II may also be a potent stimulus for an increase in production or secretion of ET in endothelial cells and cardiac fibroblasts [8, 32] (Fig. 3). Moreover, a positive feedback may exist between ANG II and ET based upon reports that ET may also enhance the conversion of ANG I to ANG II [33].

We recently explored the cause and effect relationship between the increase in ANG II and ET in CHF [16]. We tested the hypothesis that ANG II serves as a potent stimulus for increased tissue and circulating ET in CHF by chronically inhibiting ANG II generation over a one week period. Studies were performed in a model of experimen-

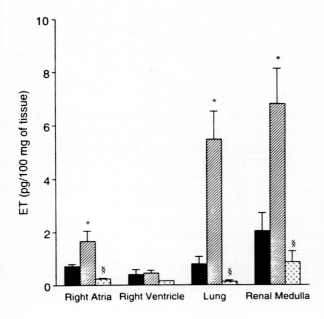

Fig. 4. Tissue ET concentrations in normal dogs $(n = 4)$ *(solid bars),* after 7 days of TIVCC $(n = 4)$ *(hatched bars),* and after 7 days of TIVCC in the presence of ACE-I $(n = 4)$ *(cross-hatched bars).* $^*p < 0.05$ vs. normals; $p < 0.05$ vs. TIVCC [16]

tal CHF produced by chronic TIVCC. This model is characterized by marked systemic and regional vasoconstriction in association with decreased cardiac output and marked activation of both the RAAS and ET systems. Chronic ACE-I resulted in a reduction in the increase in circulating ET and marked attenuation in the activation at the tissue level in the lung, heart, vasculature and kidney (Fig. 4). These studies therefore support a key role for ANG II as a participant in the activation of the ET system in experimental CHF.

Functional Importance of an Activated ET System in CHF

The functional significance of elevated circulating pathophysiologic concentrations of ET as observed in CHF was first suggested in studies by Lerman et al. [34]. In this study, exogenous ET was administered to anesthetized dogs at a concentration to increase plasma ET two-fold as observed in CHF. Pathophysiologically relevant increases in circulating ET resulted in significant systemic and renal vasoconstriction in association with a decrease in heart rate and cardiac output in the absence of an increase in arterial pressure. These studies are consistent with an important role for ET in cardio-renal regulation in CHF.

Cavero et al. extended these previous investigations utilizing a model of pacing-induced CHF. In response to a two-fold increase in baseline ET, the systemic and renal vasoconstriction was observed but the responses were attenuated compared to normal controls [27]. Possible explanations for such attenuated responses to exogenous ET in CHF include receptor down regulation related to sustained increases in ET, the presence of a preconstricted vasculature or the concomitant activation of opposing endogenous vasodilator systems. Employing the model of chronic TIVCC, Underwood et al. observed that exogenous ET in both normal and TIVCC dogs resulted in similar responses as demonstrated by full vasoconstrictor responses despite chronic increases in ET and intense systemic and renal vasoconstriction at baseline in the TIVCC group [35]. These studies suggest that neither receptor down-regulation nor diminished vaso-constrictor reserve represent dominant mechanisms accounting for reduced hemo-dynamic responses to exogenous ET in pacing induced CHF.

An important distinction between pacing-induced CHF and TIVCC is the presence of atrial stretch and marked increases in ANP in pacing-induced CHF with an absence of atrial distention or increased ANP in the preload limited TIVCC model. To define the role of ANP in modulating the biological responses to exogenous ET in TIVCC, Underwood et al. acutely administered exogenous ANP in the presence of chronic TIVCC. In this setting with pathophysiologic plasma concentrations of ANP, vascular responses to exogenous ET were markedly attenuated as was observed in pacing-induced CHF [27, 35]. This suggests an important role for ANP in counterregulating the systemic vasoconstrictor actions of ET. In contrast to the systemic circulation, ANP appeared to have little modulating effect on the potent renal vasoconstricting actions of ET in the TIVCC model. This may reflect the enhanced sensitivity to ET present in the renal vascular bed, compared with other vascular beds [6] or a dominance of ET actions within the renal circulation [36].

Endothelium-derived relaxing factor (EDRF), presumed to be nitric oxide (NO), is another important endogenous vasodilator [37, 38, 39] which may functionally attenuate responses to ET in CHF. Employing the competitive inhibitor of EDRF, NG-monomethyl-L-arginine (L-NMMA), studies indicate that EDRF contributes to the control of basal systemic and regional vascular resistances. To assess the functional antagonism between EDRF and ET responses, Lerman et al. observed marked potentiation of the systemic, renal, pulmonary and coronary arterial responses to low dose ET [40]. These studies are consistent with the hypothesis that the endogenous EDRF system serves as a functionally important modulator of the vasoconstrictor actions of ET. Additionally, these latter studies may also be interpreted as suggesting the ET-B-mediated release of EDRF during ET administration may limit the magnitude of ET mediated vasoconstriction.

Investigations suggest that basal EDRF generation may be preserved [41] in the setting of chronic CHF. From this perspective, intact or enhanced EDRF activity, like ANP, may contribute to attenuated ET-mediated vasoconstriction in CHF. Moreover, the balance between these endothelial-derived vasoconstrictor and vasodilator factors may represent an important mechanism of altered vascular tone and responses characteristic of CHF.

From the above discussion, it is clear that ET at pathophysiologic concentrations has biological actions in the control of vascular tone. Recent studies would also suggest that the predominant receptor subtype mediating the systemic and regional vasoconstrictor response to ET at pathophysiologic doses is the ET-A receptor both in the renal and coronary circulations [36, 42] (Fig. 5). Moreover, this action appears applicable to experimental CHF. In the CHF model produced by TIVCC, systemic administration of a selective ET-A receptor antagonist resulted in systemic vasodilatation and a decrease in arterial pressure [16] (Fig. 6). This is consistent with a role for endogenous ET via the ET-A receptor in maintaining arterial pressure and mediating vasoconstriction in CHF. Utilizing a mixed ET-A and ET-B receptor antagonist, such findings have also been observed in human CHF [43].

A vasoconstrictor role has also been reported for the ET-B receptor in selective vascular beds. Such an action may be enhanced in CHF. In the TIVCC model of

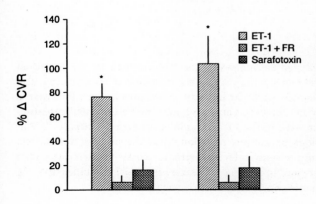

Fig. 5. Percentage change (%) in coronary vascular resistance *(CVR)* compared with baseline in three experimental groups. Group 1: *ET-1,* endothelin-1 plus vehicle; Group 2: *ET-1 + FR,* endothelin-1 plus endothelin-A blocker (FR-139317); Group 3: *Sarafotoxin,* sarafotoxin plus vehicle; substances were infused at low dose concentrations for 2 consecutive 20 min. *$p < .01$ between the groups [42]

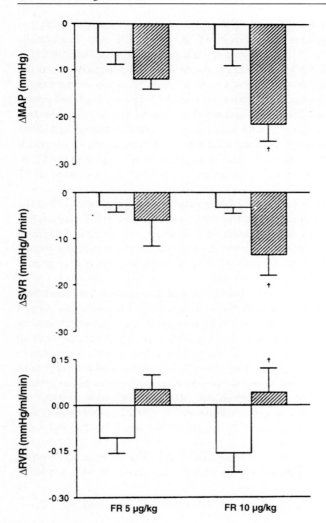

Fig. 6. Absolute changes from baseline in MAP (ΔMAP), SVR (ΔSVR), and RVR (ΔRVR) in controls (group II) *(solid bars)* and after 7 days of TIVCC (group III) *(hatched bars)* during acute intravenous FR-139317 administration at 5 and 10 µg/kg per minute. $p < 0.05$ vs. controls [16]

congestive failure, ET-B mediated coronary vasoconstriction with the ET-B agonist Sarafatoxin-6-c (S6c) was enhanced [44]. More recently, S6c administration also resulted in an exaggerated increase in forearm vascular resistance when administered intra-arterially to humans with CHF [45]. Thus, ET-B mediated vasoconstriction may be a more important characteristic of the ET system in vascular regulation in CHF as compared to normal physiologic conditions. Additionally, studies suggest that the ET-B receptor may participate in the ventricular hypertrophy which characterized overt CHF, yet data conflicts with regard to a firm conclusion [46]. Further studies utilizing specific ET-B receptor antagonists in CHF are clearly warranted.

Diagnostic and Prognostic Significance of Increased Circulating ET in CHF

The elevation of circulating ET in CHF may also have diagnostic and prognostic implications. In the study of Margulies and co-workers from our laboratory, despite overlap, the calculated sensitivity (85.7%) and specificity (82.1%) of plasma ET as a predictor of CHF were high [13]. Thus, circulating ET may serve as an important diagnostic serum marker for CHF. Such a concept is supported by the investigations of Rodeheffer et al. in humans with chronic CHF in which circulating ET increased progressively with the severity of CHF [47]. More recently, plasma ET determination at the time of acute myocardial infarction was a strong and independent predictor of mortality at one year following infarction [48].

The prognostic significance for plasma ET was recently reported by Galatius-Jensen et al. [49]. Forty-four patients with CHF showed increased plasma ET concentrations approximately twofold higher than 21 healthy volunteers. Plasma ET increased with the severity of the disease. All patients were followed for an average of seventeen months. Mortality was 27% in the group with plasma ET concentrations above 3 pg/ml and 0% in the group with plasma ET concentrations below 3 pg/ml. In contrast, separation of patients based on NYHA classification and ejection fraction revealed no significant difference regarding mortality. The authors concluded that increased plasma ET concentrations in patients with CHF appears to serve as a predictor of cardiac death.

Future Areas of Research and Therapeutics

Based upon the evidence that ET may mediate vasoconstriction and possible ventricular remodeling which characterizes CHF, one may conclude that it may be beneficial to antagonize the effects of ET in heart failure. Indeed, various strategies now are emerging. Recognizing that a unique endothelin converting enzyme (ECE) may exist which processes Big ET to the mature peptide, an inhibitor of ECE could serve to inhibit production of ET. Decreasing stimuli to ET secretion would be another possible mechanism to antagonize this peptide system. This could be accomplished by decreasing ANG II concentrations via ACE-I, angiotensin receptor antagonism or renin inhibition. Since the development of several specific endothelin receptor antagonists [50], numerous studies have demonstrated beneficial acute hemodynamic effects in models of CHF as well as in human CHF. Thus, chronic endothelin receptor antagonism may emerge as an important new strategy in the treatment of CHF. Indeed, in a most recent study, 3-month treatment of the rat infarct model of CHF with an ET-A selective antagonist resulted in a marked improvement in survival [51].

Potentiation of ANP, a known inhibitor of ET release via cGMP, with an inhibitor of ANP degradation via blockade of neutral endopeptidase 24.11 could retard ET activation in CHF [52]. In addition, the vascular component of the natriuretic peptide family, C-type natriuretic peptide (CNP) which activates cGMP and is both vasoactive and growth-inhibiting, could also serve therapeutically as a local and potent inhibitor of the ET system [53]. In addition, nitrates mediate in part their actions through stimulating the same intracellular messenger, cGMP, as does EDRF and ANP. Thus, some of the currently available treatment strategies for CHF likely have some of their beneficial effects by counteracting the actions of ET [54].

In summary, ET is a potent vasoconstrictor and mitogenic peptide which is elevated in CHF and likely contributes to the observed vasoconstriction and possible ventricular remodeling in this pathophysiologic state. This action serves as a biological paradigm of its physiologic role to preserve arterial pressure in response to an underfilling of effective arterial volume. It is likely that therapeutic regimens aimed at countering the potent vasoconstricting and potentially cardiac remodeling effects of ET will emerge as an important new strategy in the management of CHF.

Acknowledgments. This work was supported by grants from the National Institute of Health (HL-36634), the Hearst Foundation, Mayo Foundation and the Miami Heart Research Institute. DB was supported by an NIH Training Grant in Cardiovascular Disease (HL-07111).

References

1. Yanagisawa M, Kurihara H, Kimura, Tomobe Y, Kobayashi M, Mitsui Y, Yazaki K, Goto K, Masaki T (1988) A novel potent vasoconstrictor peptide produced by vascular endothelial cells. Nature 332:411–415
2. Miller VM, Komori K. Burnett JC Jr, Vanhoutte PM (1989) Differential sensitivity to endothelin in canine aneries and reins. Am J Physiol 257:H1127–H1131
3. Arai H, Hori S, Aramori H, Ohkubo H, Nakamishi S (1990) Cloning and expression of a cDNA encoding an endothelin receptor. Nature 348:732–735
4. Ito H, Hirata Y, Hiroe M, Tsujino M, Adachi S, Takamoto T, Nitta M, Taniguchi K, Marumo F (1991) Endothelin-1 induces hypertrophy with enhanced expression of muscle-specific genes in cultured neonatal rat cardiomyocytes. Circ Res 69:209–215
5. Neuser D, Knorr A, Stasch J, Kazda S (1990) Mitogenic activity of endothelin-1 and -3 on vascular smooth muscle cells is inhibited by atrial natriuretic peptides. Artery 17:311–324
6. Miller WL, Redfield MM, Burnett JC Jr (1989) Integrated cardiac, renal, and endocrine action of endothelin. J Clin Invest 83:317–320
7. Goetz KL, Wang BC, Nadwed JB, Zhu JL, Leadley RJ Jr (l988) Cardiovascular, renal, and endocrine responses to intravenous endothelin in conscious dogs. Am J Physiol 255:R1064–R1068
8. Fujisaki H, Ito H, Hirata Y, Tanaka M, Hata M, Lin M, Adachi S, Akimoto H, Marumo F, Hiroe M (1995) Natriuretic peptides inhibit angiotensin II-induced proliferation of rat cardiac fibroblasts by blocking endothelin-1 gene expression. J Clin Invest 96:1059–1065
9. Masaki T (1993) Overview: reduced sensitivity of vascular response to endothelin. Circulation 87 (Suppl V):V33–V35
10. Cernacek P, Stewart DJ (1989) Immunoreactive endothelin in human plasma: Marked elevation in patients in cardiogenic shock. Biochem Biophys Res Commun 161:562–567
11. Saito Y, Nakao K, Mukoyama M, Imura H (1990) Increased plasma endothelin level in patients with essential hypertension (letter) N Engl J Med 322:205
12. Wei CM, Lerman A, Rodeheffer RJ, McGregor CGA, Brandt RR, Wright S, Heublein DM, Kao PC, Edwards WD, Burnett JC (1994) Endothelin in human congestive heart failure. Circulation 89:1580–1586
13. Margulies KB, Hildebrand FL Jr, Lerman A, Perrella MA, Burnett JC Jr (1990) Increased endothelin in experimental heart failure. Circulation 82:2226–2230
14. Lerman A, Click RL, Narr BJ, Weisner RH, Krom RA, Textor SC, Burnett JC Jr (1991) Elevation of plasma endothelin associated wiith systemic hypertension in humans following orthotopic liver transplantation. Transplantation 51:646–650

15. Margulies KB, Hildebrand FL, Heublein DM, Burnett JC Jr (1991) Radiocontrast increases plasma and urine endothelin. J Am Soc Nephrol 2:1041–1045
16. Clavell AL, Mattingly MT, Nir A, Aarhus LL, Heublein DM, Burnett JC Jr (1996) Angiotensin converting enzyme inhibition modulates endogenous endothelin in chronic canine thoracic inferior vena caval constriction. J Clin Invest 97:1286–1292
17. Lerman A, Edwards BS, Hallett JW, Heublein DM, Sandberg SM, Burnett JC Jr (1991) Circulating and tissue endothelin immunoreactivity in advanced atherosclerosis. New Engl J Med 325:997–1001
18. Ravalli S, Szabolcs M, Albala A, Michler RE, Cannon PJ (1996) Increased immunoreactive endothlin-1 in human transplant coronary artery disease. Circ 94:2096–2102
19. Nunez DJR, Brown MJ, Davenport AP, Neylon CB, Schofield JP, Wyse RK (1990) Endothelin-1 mRNA is widely expressed in porcine and human tissues. J Clin Invest 85:1537–1541
20. Saito T, Yanagisawa M, Miyauchi T, Suzuki N, Matsumoto H, Fujino M, Masaki T (1989) Endothelin in human circulating blood: effects of major surgical stress (abstract). Jpn J Pharmacol 49:2l5P
21. Wei CM, Clavell AL, Burnett JC Jr. Increased cardiopulmonary endothelin gene expression in experimental heart failure. Am J Physiol (in press)
22. Edwards BS, Zimmerman RS, Schwab TR, Heublein DM, Burnett JC Jr (1988) Atrial stretch, not pressure, is the principal determinant controlling the acute release of atrial natriuretic factor. Circ Res 3:191–195
23. Cody RJ, Haas GJ, Binkley PF, Capers Q, Kelley R (1992) Plasma endothelin correlates with the extent of pulmonary hypertension in patients with chronic congestive heart failure. Circulation 85:504–509
24. Kourembanas S, Marsden PA, Mcquillan LP, Faller DV (1991) Hypoxia induces endothelin gene expression and secretion in cultured human endothelium.J Clin Invest 88:1054–1057
25. Nir A, Clavell AL, Heublein D, Aarhus L, Burnett, JC, Jr (1994) Acute hypoxia and endogenous renal endothelin. J Am Soc Nephrol 4:1920–1924
26. Sandok EK, Lerman A, Stingo A, Perrella MA, Gloviczki P, Burnett JC Jr (1992) Endothelin in a model of acute renal dysfunction: modulating action of atrial natriuretic factor. Am Soc Nephrol 3:196–202
27. Cavero P, Miller WL, Heublein DM, Margulies KB, Burnett JC Jr (1990) Endothelin in experimental congestive heart failure in the anesthetized dog. Am J Physiol 259:F312–F317
28. Loeffler BM, Roux S, Kalina B, Clozel M, Clozel JP (1993) Influence of congestive heart failure on endothelin levels and receptors in rabbits. J Molec Cardiol 25:407–416
29. Fukuroda T, Fujikawa T, Ozaki S. Ishikawa K, Yano M, Nishikibe M (1994) Clearance of circulating endothelin-1 by ET_B receptor in rats. Biochem Biophys Res Commun 199 (3):1461–1465
30. SAVE Investigators (1992) Effect of captopril on mortality and morbidity in patients with left ventricular dysfunction after myocardial infarction. N Engl J Med 327:669–677
31. SOLVD Investigators (1991) Effect of enalapril on survival in patients with reduced left ventricular ejection fractions and congestive heart failure. N Engl J Med 325:293–302
32. Kohno M, Horio T, Ikeda M, Yokokawa K, Fukui T, Yasunari K, Murakawa K, Kurihara N, Takeda T (1993) Natriuretic peptides inhibit mesangial cell production of endothelin induces by arginine vasopressin. Am J Physiol 264:F678–F683
33. Kawaguchi H, Sawa H, Yasuda H (1990) Endothelin stimulates angiotensin I to angiotensin II conversion in cultured pulmonary artery endothelial cells. J Mole Cell Cardiol 22:839–842
34. Lerman A, Hildebrand FL Jr, Aarhus LL, Burnett JC Jr (1991) Endothelin has biological actions at pathophysiological concentrations. Circulation 83:1808–18l4
35. Underwood RD, Aarhus LL, Heublein DM, Burnett JC Jr (1992) Endothelin in thoracic inferior vena caval constriction. Am J Physiol 263 (32):H951–H955
36. Clavell AL, Stingo AJ, Margulies KB, Brandt RR, Burnett JC Jr (1995) The role of endothelin receptor subtypes in the in-vivo regulation or renal function. Am J Physiol (Renal, Fluid and Electrolye Physiol) 268:F455
37. Furchgott RF, Zawadski JV (1980) The obligatory role of the endothelial cells in the relaxation of arterial smooth muscle by acetylcholine. Nature 188:373–376
38. Ignarro LJ, Buga GM, Wood KS, Byrns RE (1987) Endothelium-derived relaxing factor produced and released from artery and vein is nitric oxide. Proc Natl Acad Sci USA 84:9265–9269
39. Palmer RMJ, Ashton DS, Moncada S (1988) L-Arginine is the physiological precursor for the formation of nitric oxide in endothelium-dependent relaxation. Biochem Biophys Res Commun 53:1251–1256
40. Lerman A, Sandok EK, Hildebrand FL Jr, Burnett JC Jr (1992) Inhibition of endothelium-derived relaxing factor enhances endothelin-mediated vasoconstriction. Circulation 85:1894–1898

41. Drexler H, Lu W (1992) Endothelial dysfunction of hindquarter resistance vessels in experimental human failure. Am J Physiol 262:H1640–H1645
42. Cannon CR, Burnett JC Jr., Brandt RR, Lerman A (1995), Endothelin at pathophysiological concentrations mediates coronary vasoconstriction via the endothelin-A receptor. Circulation 92:3312–3317
43. Kiowski W, Stutsch G, Hunziker P, Muller P, Kim J, Oechslin E, Schmitt R, Jones R, Bertel O (1995) Evidence for endothelin-1 mediated vasoconstriction in severe chronic heart failure. Lancet 346:732–736
44. Cannon CR, Burnett JC Jr., Lerman A (1996) Enhanced coronary vasoconstriction to endothelin-B-receptor activation in experimental congestive heart failure. Circulation 93:646–651
45. Love MP, Haynes WG, Gray GA, Webb DJ, McMurray JJV (1996) Vasodilator effects of endothelin-converting enzyme inhibition and endothelin ET-A receptor blockade in chronic heart failure patients treated with ACE inhibitors. Circulation 94:2131–2137
46. Tamamori M, Ito H, Adachi S, Akimoto H, Marumo F, Hiroe M (1996) Endothelin-3 induces hypertrophy of cardiomyocytes by the endogenous endothelin-1-mediated mechanism. J Clin Invest 97:366–372
47. Rodeheffer RJ, Lerman A, Heublein DM, Burnett JC, Jr (1992) Increased plasma concentrations of endothelin in congestive heart failure. Mayo Clin Proc 67:719–724
48. Omland T, Lie RT, Aakvaag A, Aarsland T, Dickstein K (1994) Plasma endothelin determination as a prognostic indicator of 1-year mortality after acute myocardial infarction. Circulation 89:1573–1579
49. Galatius-Jensen S, Wroblewski H, Emmeluth C, Bie P, Haunso S, Kastrup J: Plasma endothelin in congestive heart: a predictor of cardiac death (1996) J Cardiac Failure 2:71–75
50. Opgenorth TJ, Adler AL, Calzadilla SV, Chiou WJ, Dayton BD, Dixon DB, Gehrke LJ, Hernandez L, Magnuson SR, Marsh KC, Novosad EI, Von Geldern TW, Wessale JL, Winn M, Wu-Wong JR (1996) Pharmacological characterization of A-127722: an orally active and highly potent ET-A selective receptor antagonist. J Pharmacol Exp Therap 276:473–481
51. Sakai S, Miyauchi T, Kobayashi M, Yamaguchi I, Goto K, Sugishita Y (1996) Inhibition of myocardial endothelin pathway improves long-term survival in heart failure. Nature 384:353–355
52. Margulies KB, Barclay PL, Burnett JC Jr (1995) The role of neutral endopeptidase in dogs with evolving congestive heart failure. Circulation 91:2036–2042
53. Kullo IJ, Burnett, JC Jr (1996) C-type Natruiretic Peptide: The Vascular Component of the Natriuretic Peptide System. In Contemporary Endocrinology: Endocrinology of the Vasculature, Humana Press Inc., Totowa NJ, pp 9–93
54. Miller WL, Cavero PG, Aarhus LL, Heublein DM, Burnett JC Jr (1993) Endothelin-mediated cardiorenal hemodynamics and neuroendocrine effects are attenuated by nitroglycerin in vivo. Am J Hypertens 6:156–163

Endothelin in Human Cardiovascular Physiology and Pathophysiology

D. J. Webb

Introduction

Endothelin-1 was first identified by Yanagisawa and his colleagues in 1988 from the supernatant of endothelial cells in culture [1] and shown to be an extremely potent vasoconstrictor and pressor peptide. It is now known that endothelial cells, which line all blood vessels, are capable of generating endothelin-1 and that receptors for the endothelins are widely expressed, particularly in tissues involved in cardiovascular regulation, including the heart, blood vessels, kidney and brain. Endothelin-1 has potent direct vasoconstrictor properties – to which the coronary, renal and cerebral blood vessels appear particularly sensitive – and the capacity to enhance vasoconstriction to other agents. In addition, endothelin-1 has major activity as a co-mitogen. Taken together, these properties indicate a likelihood that the endothelin system is of functional importance in human cardiovascular physiology, and have attracted major attention to the possibility that endothelin-1 may play a role in the pathophysiology of cardiovascular disease and, hence, that endothelin antagonists may prove useful therapeutic agents.

While studies in animals indicating the involvement of endothelin-1 in the pathophysiology of heart failure and some forms of hypertension are compelling, the evidence in animals for a physiological role of the endothelin system in maintenance of blood pressure has been less clearcut (see [2]). However, there are studies showing that ECE inhibitors and endothelin receptor antagonists can reduce blood pressure in normotensive animals [3–5], usually where haemodynamic responses have been followed for several hours after drug administration. Nevertheless, it is clear that there are important species differences between rat, dog and human: including the anatomical localisation of the receptors; the isoforms of endothelin found; and the receptors subtypes mediating vascular and renal effects, so the results of studies in experimental animals cannot necessarily be extrapolated to man.

Several requirements are necessary before it is reasonable to attribute a role to the endothelin system in human cardiovascular disease. These include: evidence that endothelin-1 mimics, or promotes, the consequences of the disease; that the production of endothelin-1 is increased or its clearance decreased, or that the number or sensitivity of endothelin receptors is increased in the disease; and that inhibitors of the production or actions of the endothelins ameliorate the disease process. In many cardiovascular conditions some of these requirements have been fulfilled (Table 1) but, in most,

Table 1. Important cardiovascular diseases in which endothelin-1 has been implicated (For detailed review see reference [39])

Sustained vasoconstriction
  Essential hypertension
  Chronic heart failure
  Chronic renal failure
  Primary pulmonary hypertension
  Cyclosporine hypertension and nephrotoxicity

Intermittent vasoconstriction
  Unstable angina
  Acute renal failure
  Subarachnoid haemorrhage
  Raynaud's disease
  Migraine

not all are currently satisfied. Plasma endothelin concentrations may be misleading for a number of reasons: first, unless the controls are well matched in all respects a high plasma endothelin concentration in patients may be spurious; second, endothelin-1 is released abluminally and is not generally considered to be a circulating hormone, so the absence of an increase in plasma endothelin concentration does not necessarily exclude increased production, particularly given the relative imprecision of most of the assays. Vascular sensitivity to endothelin-1 may also be misleading because increased production may lead to receptor downregulation and therefore reduced sensitivity to exogenous administration of the peptide may indicate *overactivity* of the system, underlining the importance of studies with endothelin antagonists.

The recent availability of inhibitors of the synthesis and actions of endothelin for clinical use has substantially contributed to our understanding of the role of endothelin-1 in cardiovascular physiology and pathophysiology. As yet, however, few studies with these drugs have been published. Nevertheless, the development of endothelin receptor antagonists has become a very competitive field within the pharmaceutical industry and a number of drugs are now entering large-scale clinical trials. It is likely, therefore, that our understanding of the field, and the therapeutic benefits that may be associated with the use of these agents, will emerge more clearly over the next few years. The aim of this final chapter in the section on the endothelin system is to review two major clinical areas, hypertension and heart failure, in which some of the best evidence for benefit currently exists.

Human Cardiovascular Physiology

Endothelin-1 is currently the most potent vasoconstrictor and pressor agent known [1]. Exogenous endothelin-1 is a pressor agent in man [6], causing also reductions in renal plasma flow, glomerular filtration rate and urine sodium excretion. None of these effects are seen with administration of equimolar doses of endothelin-3, suggesting that these effects are mediated predominantly through the ET_A receptor. Given locally into regional circulations, endothelin-1 causes uniquely sustained constriction of forearm resistance vessels and hand veins [7], as well as of dermal microvessels [8]. A potential contribution to vasoconstriction in humans from vascular ET_B receptors is suggested by the vasoconstriction produced by relatively selective ET_B receptor agonists, such as endothelin-3 and sarafotoxin S6c [9], although the former involved relatively high concentrations that may have generated effects on ET_A receptors. Vasodilatation appears to occur only with high "bolus" doses of endothelin-1 and endothelin-3 [9] and is then only transient, converting to vasoconstriction within 5 min for endothelin-1 and 15 min for endothelin-3 and, therefore, vasodilatation is unlikely be of physiological significance. The dilator response, at least in resistance vessels, appears to be resistant to aspirin and probably mediated by nitric oxide release. The sustained vasoconstriction and venoconstriction to exogenous endothelin-1 appear to be modulated by a counterbalancing release of endothelium dependent dilators; nitric oxide in the resistance vessels and dilator prostaglandins in veins [10], presumably through an action on the endothelial cell ET_B receptor. In studies in resistance vessels [11] and hand veins [12], there was no evidence that local administration of endothelin-1 could amplify peripheral sympathetic responses, as has been shown in vitro [13].

The significance of endothelin-1 in maintenance of resistance vessel tone has been clarified with studies using selective and non-selective endothelin receptor antagonists and endothelin converting enzyme (ECE) inhibitors. In the forearm resistance vessels [14], using local brachial artery administration, the ECE inhibitor, phosphoramidon, and the ET_A receptor antagonist, BQ-123, inhibition of endogenous endothelin-1 generation and its actions at the ET_A receptor have been shown to produce progressive and substantial vasodilatation (Fig. 1). These findings show that endothelin-1 makes a major contribution to the maintenance of basal resistance vessel tone in healthy human subjects and suggest that activation of the vascular smooth muscle ET_A receptor provides the major component of endothelin-1 mediated tone. This also appears to be the case for the dermal microvessels, because vasoconstriction to endothelin-1 can be abolished by co-administration of an ET_A selective receptor antagonist [8]. Interestingly, more recent studies show that arterial administration of the ET_B selective receptor antagonist, BQ-788, causes forearm vasoconstriction [15], consistent with the major physiological role of ET_B receptor stimulation being vasodilatation, i.e. under physiological conditions, actions of endothelin-1 at the endothelial ET_B receptor outweigh those on the vascular smooth muscle ET_B receptor. This is also consistent with the observation that brachial artery administration of the ET_A selective receptor antagonist, BQ-123, causes substantially greater vasodilatation of the forearm vessels than administration of the mixed $ET_{A/B}$ receptor antagonist, TAK-044 [16]. Other recent but as yet unpublished studies suggest that a component of the vasodilatation to BQ-123 is sensitive to BQ-788 or to the nitric oxide synthase inhibitor, L-N^G-monomethyl arginine (LNMMA), implying that the response is at least in part an indirect result of enhanced endothelial ET_B receptor mediated nitric oxide induced vasodilatation as well as directly to a withdrawal of ET_A receptor mediated tone. In the studies with phosphoramidon, plasma concentrations of big endothelin-1 and of the two products of big endothelin-1 cleavage, endothelin-1 and the C-terminal fragment, were separately measured using HPLC separation followed by radioimmunoassay using antibodies

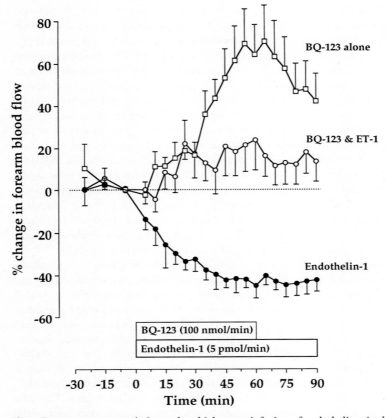

Fig. 1. Forearm vasoconstriction to brachial artery infusion of endothelin-1 is abolished by the coinfusion of BQ-123. Infusion of BQ-123 alone produces progressive forearm vasodilatation. (From [14]) with permission

selective for the C- and N-terminal segments of big endothelin-1 [17]. The results indicate that plasma big endothelin-1 may give a better indication of endothelin-1 generation than plasma endothelin-1, because the latter is very rapidly cleared from the circulation, presumably by binding to receptors in vascular tissue. Plasma C-terminal fragment concentrations seem less reliable because this peptide fragment has a relatively short plasma half-life [18].

Responses of the forearm resistance vessels to drugs given into the brachial artery are usually predictive of the responses in the major resistance beds that serve to regulate blood pressure [19] so it might be envisaged that endothelin-1 plays a role in maintenance of systemic vascular resistance. This has been shown using a potent peptide agent, the mixed $ET_{A/B}$ receptor antagonist, TAK-044. Given systemically to healthy people in a dose ranging study, a 15 min infusion of TAK-044 (10–1000 mg) reduced systemic vascular resistance in a dose-dependent manner [16]. At 1000 mg, TAK-044

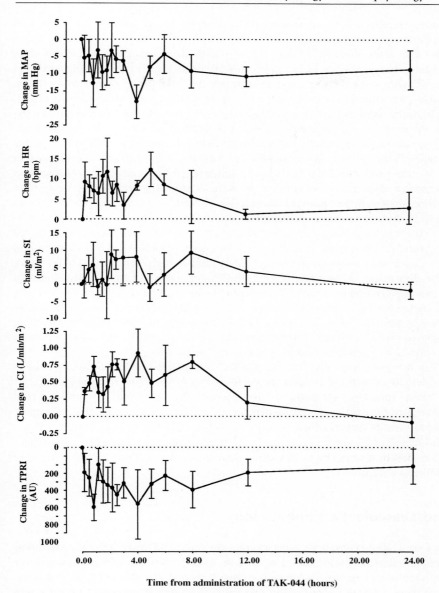

Fig. 2. Time course of the effects of the highest dose of TAK-044 (1000 mg) on mean arterial pressure *(MAP)*, heart rate *(HR)*, stroke index *(SI)*, cardiac index *(CI)*, and total peripheral resistance index *(TPRI)*. TAK-044 significantly decreased mean arterial pressure ($p < 0.001$) and total peripheral resistance ($p < 0.001$) and increased heart rate ($p < 0.001$), stroke index ($p = 0.034$) and cardiac index ($p < 0.001$); these effects were maximal at 4 h and sustained for at least 12 h. Data shown represent placebo-corrected changes from predose [change from predose (active) – mean change from predose (placebo)]. *bpm*, beats per minute; *AU*, arbitrary units. (From [16]) with permission

reduced systolic blood pressure by ~4%, diastolic blood pressure by ~18%, and systemic vascular resistance by ~26% over a 24-h period (Fig. 2), suggesting that a major part of the effect of TAK-044 is mediated in the resistance vessels. At this dose, the haemodynamic effects were accompanied by increases in cardiac output and heart rate, although these compensatory effects were not consistently seen at lower doses. Similar observations have been made with the mixed $ET_{A/B}$ receptor antagonist, bosentan, in health subjects [20]. Here, an approximately 5 mm Hg reduction in blood pressure was seen, in association with a 5 beat/min increase in heart rate, and again effects on heart rate were apparently not consistent although haemodynamic effects in this study were reported only in summary form. These results provide strong evidence that endogenous production of endothelin-1 is important in the maintenance of basal vascular tone and in the control of blood pressure, as originally proposed by Yanagisawa [1]. Nevertheless, these findings remain to be confirmed by other workers and extended to studies with other endothelin receptor antagonists. In this regard, it might be predicted that ET_A selective receptor antagonists would be more effective than mixed $ET_{A/B}$ receptor antagonists and this clearly needs confirmation because of its clinical implications in hypertension and heart failure. Interestingly, and perhaps of relevance to the clinical use of these agents, systemically administered TAK-044 also abolished the vasoconstriction to endothelin-1 infused via the brachial artery for up to 3 h [16], although endothelin-1 induced vasoconstriction substantially recovered by 8 h after dosing. A sustained ability to block endothelin mediated "vasospasm" rather than simply to cause sustained systemic haemodynamic actions might be of particular importance in conditions such as subarachnoid haemorrhage.

Now that endothelin receptor antagonists are available for clinical use it will be of considerable interest and importance to address more closely the effects of these agents on renal function, sympathetic nervous activity and other neurohormonal mechanisms for cardiovascular homeostasis. Furthermore, it will be of some importance to identify factors which alter the regulation of the endothelin system – in the blood vessels or kidney, for example – in order to predict how different manoeuvres – such as diet or diuretic induced sodium depletion – may interact with the effects of endothelin antagonists.

Cardiovascular Pathophysiology

There are a number of mechanisms whereby the activity of the endothelin system may be enhanced in cardiovascular disease (Table 2) and a number of ways in which the administration of endothelin antagonists may be broadly beneficial (Table 3). A number of these issues are addressed in relation to the pathophysiology of hypertension and chronic heart failure.

Hypertension

Essential hypertension is a condition which, when established, is characterised by a high blood blood pressure in association with an increased peripheral vascular

Table 2. Mechanisms whereby activity of the endothelin system may be enhanced (For detailed review see [39])

Increased endothelin production
Increased sensitivity or responsiveness to endothelin
Reduced endothelin clearance (mainly renal and pulmonary)
Sustained endothelin activity in the presence of impaired dilator function

Table 3. Mechanisms whereby endothelin antagonists may produce benefit in cardiovascular disease (For detailed review see [39])

Arteriolar vasodilatation
Dermal microvascular vasodilatation
Venodilatation
Coronary vasodilatation
Sympatholytic activity
Increased urine sodium excretion
Regression/prevention of atherogenesis
Regression/prevention of cardiac and vascular hypertrophy
Prevention of maladaptive remodelling after myocardial infarction

resistance. In the early stages there may be enhanced activity of the sympathetic nervous system. On the basis of the initial pharmacology described in their original paper, Yanagisawa and colleagues suggested that disturbances in the control of endothelin production might contribute to the pathogenesis of hypertension [1]. This hypothesis was extended further with the evidence that endothelin-1 is a co-mitogen, enhancing cell division and proliferation, gene expression, protein synthesis and, ultimately, promoting hypertrophy of vascular smooth muscle as well as cardiac myocytes and fibroblasts [21–23]. Additionally, it is now recognised that endothelin-1 can exert indirect effects on vascular tone: by augmenting vasoconstriction to other agents such as angiotensin II [24], noradrenaline and serotonin [13]; by enhancing central [25] and peripheral sympathetic function [13] and by activating the renin-angiotensin system [26]. Hence, endothelin-1 might be a direct cause of hypertension or act indirectly to amplify the effects of other mediators. In order to sustain hypertension, there may be a requirement for influences on renal sodium handling as well as vascular and cardiac effects, for which there is already some evidence in animals [27, 28]. Endothelial dysfunction relating to the L-arginine-nitric oxide system in cardiovascular disease, including hypertension [29], is already well recognised. Therefore, even if impaired endothelial function is confined to the nitric oxide system, the balance between endothelium-derived dilator and constrictor factors may alter to favour vasoconstriction if the endothelin system were not also to be downregulated, i.e. a fine balance between the two systems needs to be maintained. Abnormalities of the endothelin system in severely hypertensive animals are now well described (see Schiffrin, this volume) though the evidence in humans is still incomplete.

Early investigators [30] detected elevated concentrations of circulating immunoreactive endothelin in severely hypertensive patients and proposed that increased production of endothelin-1 might cause essential hypertension. However, it has subsequently emerged that clearance of endothelin-1 is dependent on renal function; indeed, in animal models of accelerated hypertension, plasma endothelin concentrations correlate positively with plasma creatinine [31]. Other studies, in essential hypertensives with normal renal function, have shown similar concentrations of endothelin-1 in the hypertensive patients to those in matched normotensive subjects [12, 32]. Additionally, a negative correlation between blood pressure and plasma endothelin-1 in the hypertensive group in one of these studies [32] makes it unlikely that a global increase in endothelin-1 generation contributes to essential hypertension. However, there remains the possibility of a subgroup of patients in whom increased endothelin production is important. Here, studies designed to relate genotype to endothelin phenotypes may be helpful.

Increased production of endothelin-1 may be associated with at least one secondary form of hypertension, albeit extremely rare. Yokokawa and colleagues [33] have described two cases of the skin tumour, haemangio-endothelioma, in which hypertension was associated with increased plasma endothelin concentrations. Biopsies of tumour cells displayed increased expression of mRNA for endothelin-1 and strong immunohistochemical staining for the peptide. Blood pressure and plasma endothelin-1 returned to normal in both cases following surgical resection of the tumours and, in one patient, recurrence of the tumour led to further increase in both blood pressure and plasma endothelin-1.

In studies comparing Wistar-Kyoto (WKY) and spontaneously hypertensive rats (SHR), both conduit and resistance vessels from the SHR are more sensitive in vitro to the effects of endothelin-1 [34, 35]. The mechanism for this enhanced sensitivity to endothelin-1 is unclear, as the number of binding sites for endothelin-1 in aortic smooth muscle are lower in SHR [35], and these findings may relate to altered vascular structure in established hypertension. In contrast, the response to endothelin in small subcutaneous arteries of mild essential hypertensive patients were blunted [36], failing to support an increased responsiveness in human vessels. Interestingly, however, an enhanced venoconstrictor responsiveness to endothelin-1 was found in the capacitance vessels of hypertensive subjects in vivo [12]. These vessels are not thought to develop structural changes in hypertension and do not have exaggerated responses to noradrenaline or other vasoconstrictor agents. This study also demonstrated an enhancement of sympathetic mediated vasoconstriction in the hypertensive patients, but not the normotensive controls, when exposed to endothelin-1. The venoconstrictor response to endothelin-1 correlated positively with the level of arterial pressure in the hypertensives, suggesting that enhanced sensitivity to endothelin-1, at a receptor or post-receptor level, as opposed to increased production might contribute to the pathogenesis of human essential hypertension.

Unfortunately, there are currently no full published reports of studies in which endothelin antagonists have been given to hypertensive patients. However, an initial report of a study with the orally active combined $ET_{A/B}$ receptor antagonist, bosentan, suggests that endothelin antagonists can lower blood pressure [37], though in this study only to a relatively minor degree. Given into the brachial artery, the ET_A receptor antagonist, BQ-123, produces arterial vasodilatation of a similar magnitude in healthy subjects and hypertensive patients [38] suggesting that endothelin-1 may not contribute to the enhanced peripheral resistance in essential hypertension by an action on

ET_A receptors in resistance vessels. The combined $ET_{A/B}$ antagonist, TAK-044, produced rather more convincing reduction of blood pressure, and systemic vascular resistance, in normotensive subjects [16] than did bosentan in hypertensive subjects so a full report of the bosentan study and further reports with newer agents are certainly needed to clarify the position. One additional clinical issue which needs clarification, particularly in an asymptomatic condition such as hypertension, is the frequency with which side effects are associated with the use of endothelin receptor antagonists. Probably the most frequently described side effect associated with the short term use of endothelin receptor antagonists is headache. Clearly, these drugs are arteriolar vasodilators and this symptom is often found in patients using other drugs with this action. Alternatively, the indications (see earlier) that some of the vasodilatation to ET_A receptor antagonists is nitric oxide mediated raises the possibility that this may be a "nitrate" type headache. Other issues which need to be addressed in longer term studies include whether blood pressure reduction is sustained, whether there are sustained effects on the sympathetic nervous system, and whether there are other important associated benefits such as regression of left ventricular hypertrophy. In addition, there is evidence that the endothelin system is implicated in cardiovascular diseases related to hypertension that may influence how widely endothelin antagonist drugs may be used clinically in the future. Other interesting possibilities are that the endothelin system may be of particular importance in the pathophysiology of the hypertension of chronic renal failure or cyclosporine nephrotoxicity [39].

Atherosclerosis and Ischaemic Heart Disease

Atherosclerosis is a focal intimal disease of arteries, with atherosclerotic plaques characterised by accumulation of both extracellular and intracellular lipid derived from oxidised low density lipoprotein (LDL), macrophages derived from circulating monocytes, and vascular smooth muscle proliferation and migration. Affected arteries also show a tendency to develop vasoconstriction more readily than normal arteries. The endothelin system may contribute to plaque formation because of interactions at a number of stages in their development. Oxidised LDL has been shown to increase production of endothelins from macrophages [40] and from endothelial cells [41], and endothelin-1 is chemotactic for circulating monocytes and a co-mitogen for vascular smooth muscle cells [22]. Plasma endothelin concentrations are increased in patients with symptomatic atherosclerotic disease with higher concentrations of plasma endothelin correlating significantly with the number of sites of disease involvement [42]. In this study, high concentrations of immunoreactive endothelin-1 were demonstrated in both endothelial cells and the vascular smooth muscle cells of the atherosclerotic vessels. The expression of ET_B receptors has also been found to be upregulated in the coronary arteries of patients with atherosclerotic disease compared to those patients with dilated cardiomyopathy [43]. Immunoreactive endothelin-1 concentrations are increased within active plaque tissue obtained by directional atherectomy from patients with unstable angina compared with concentrations in non-active plaque tissue obtained at revascularisation in patients with stable angina [44] suggesting that endothelin-1 may contribute to the higher degree of vasospasm found in these vessels and perhaps indirectly to the risk of plaque rupture.

Plaque rupture underlies the pathogenesis of unstable angina and myocardial infarction. Endothelin-1 may potentially contribute to the induction of myocardial ischaemia after plaque rupture through its actions on the coronary vasculature. However, it is perhaps more likely that endothelin-1 exacerbates ischaemia and reperfusion injury following its release in response to tissue hypoxia [45]. Additionally, the sensitivity of cardiac tissue to endothelin-1 appears to be enhanced following ischaemia/reperfusion [46], perhaps in part due to upregulation of endothelin receptor number [47]. Plasma endothelin-1 concentrations are increased in patients with unstable angina and non-Q wave infarction [48, 49] but not those with stable angina [50]. Higher plasma concentrations of endothelin-1 are also associated with a higher degree of haemodynamic impairment in myocardial infarction [51], and are highly predictive of a poor outcome [52] and future cardiac events [49]. Clinical studies in these areas are yet to be reported.

Chronic Heart Failure

Chronic heart failure is a common and disabling condition with a high morbidity and mortality and serious economic consequences for health care [53]. Related to hypertension, chronic heart failure is often the long term consequence of ischaemic heart disease. Neuroendocrine activation – partly in response to the reduction in cardiac output – increases concentrations of noradrenaline, angiotensin II and aldosterone, and may also produce haemodynamic effects on the circulation that further compromise cardiac function. Indeed, neuroendocrine activation is associated with a poor outcome [54]. Therefore, the current mainstay of treatment for chronic heart failure is vasodilator therapy using either angiotensin-converting enzyme (ACE) inhibitors or the combination of organic nitrates and hydralazine, both of which regimens have been shown to alleviate symptoms and prolong survival in patients with chronic heart failure. However, the morbidity and mortality from heart failure remains high and there is undoubtedly room for additional effective treatment.

The mediators associated with neurohumoral activation, as well as local tissue hypoxia, should all act to increase endothelin-1 production and the actions of endothelin-1 – vasoconstriction, mitogenesis (leading to cardiac and vascular hypertrophy), enhancement of renin-angiotensin and sympathetic nervous system activity, and promotion of renal vasoconstriction and sodium retention – are all consistent with the circulatory abnormalities found in this condition. Indeed, plasma endothelin concentrations are elevated in chronic heart failure, mainly through an increase in plasma big endothelin-1 [55], consistent with increased generation of endothelin-1 [17]. Plasma endothelin correlates closely with the degree of haemodynamic and functional impairment [56–60]. Raised plasma endothelin concentrations are associated with a worse prognosis, irrespective of the cause of the cardiac failure, and predict mortality or the need for cardiac transplantation [58]. Indeed, plasma big endothelin-1 concentration is currently the most powerful predictor of outcome [60].

The increase in plasma big endothelin-1 concentrations is consistent with an increased production of endothelin-1 in heart failure rather than reduced clearance

[17]. Therefore, endothelin-1 is likely to contribute directly to neuroendocrine activation and, hence, vasoconstriction. Indeed, it has already been shown that current therapies for chronic heart failure may act at least in part by interaction with the endothelin system. In vitro, ACE inhibitors can block the stimulated release of endothelin-1 from cultured human endothelial cells [61] and in vivo, ACE inhibitors block the endothelin-1 induced increase in forearm vascular resistance when infused locally into the brachial artery [62]. ACE inhibitors reduce plasma endothelin in some [63] but not other studies [64] in patients with heart failure. Plasma endothelin concentrations are also reduced in patients with chronic heart failure in response to carvedilol, a combined α/β adrenoceptor antagonist, and this reduction correlates with an improvement in haemodynamic parameters [65]. It is not known whether carvedilol had direct effects on the production of endothelin-1 or acts indirectly to modulate neuroendocrine activation by improving systemic haemodynamics, althought there is evidence that carvedilol can inhibit endothelin-1 generation by endothelial cells in culture.

Direct blockade of the endothelin system has also been examined. In the first published clinical study [59], in patients with severe chronic heart failure withdrawn from ACE inhibitor treatment, acute intravenous administration of the combined $ET_{A/B}$ antagonist, bosentan, increased cardiac output and reduced systemic and pulmonary vascular resistance without inducing reflex tachycardia or increasing plasma concentrations of angiotensin II or noradrenaline (Fig. 3). These studies have been extended to ACE inhibitor treated patients and show that the benefits of bosentan are sustained during chronic treatment (W. Kiowski, personal communication). More recent studies, in optimally ACE inhibited patients [66], first confirm that additional haemodynamic benefits in terms of vasodilatation occurs with local brachial artery administration of the ECE inhibitor, phosphoramidon, and the ET_A receptor antagonist, BQ-123, and second show that vasoconstriction to the ET_B receptor agonist, sarafotoxin S6c, was enhanced in patients compared to responses in healthy volunteers whereas vasoconstriction to the combined $ET_{A/B}$ receptor agonist, endothelin-1, was impaired. These observations, also seen in the coronary vessels in experimental chronic heart failure [67], are consistent with upregulation of the constrictor ET_B receptor in this condition and merit further examination because of their potential clinical implications. However, importantly, both in healthy subjects and in patients with chronic heart failure, the effect of brachial artery infusion of a selective ET_B receptor antagonist, BQ-788, was to cause forearm vasoconstriction [15]. These observations imply that prevailing balance of vascular and endothelial ET_B receptor activation in resistance vessels favours endothelium dependent dilatation, even in patients with heart failure, suggesting – at least in terms of vasodilatation – that a selective ET_A receptor antagonist might be more effective than an agent with combined $ET_{A/B}$ receptor antagonist properties.

The proven reduction of mortality from heart failure by ACE inhibitors [51] was predicted from animal models [66]. Therefore, the recent report that the ET_A receptor antagonist, BQ-123, substantially improved 12-week survival from 43% to 85% in a coronary occlusion model of heart failure [69] as well as improving haemodynamic function and causing substantial amelioration of cardiac remodelling is, therefore, very promising for future clinical developments in this area. Endothelin receptor anta-

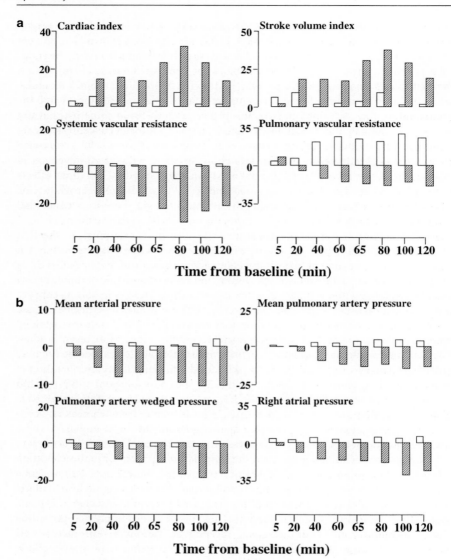

Fig. 3a, b. Change of cardiac and stroke volume index and of systemic and pulmonary vascular resistance (**a**) and changes of arterial, pulmonary artery, pulmonary artery wedged, and right atrial pressures (**b**) in patients with severe congestive heart failure after intravenous placebo (*unshaded columns*) or bosentan (*shaded columns*). Bosentan 100 mg was given intravenously at 0 min and a further 200 mg was given at 60 min. All parameters were significantly improved by acute administration of bosentan ($p < 0.05$) without change in heart rate. (From [59])

gonists are currently entering large scale trials in heart failure and it should soon be clear whether their major promise can be fulfilled under rigorous testing.

Conclusions

Endothelin-1 is the most potent and sustained vasocontrictor substance known and its discovery prompted the rapid development of endothelin converting enzyme inhibitors and endothelin receptor antagonists. A number of these endothelin antagonists are now in various stages of clinical development for the treatment of cardiovascular disease.

Using such agents as tools to explore physiological mechanisms it now appears, in contrast to some animal species, that endothelin-1 plays an important role in the maintenance of basal cardiovascular tone and blood pressure in healthy man. Additional actions, for instance on renal function and mitogenesis, and interactions with the sympathetic nervous and renin-angiotensin systems – detected during preclinical studies – remain to be confirmed in man. These physiological studies predicted the possibility that endothelin antagonists might be effective in the treatment of diseases associated with intermittent or sustained vasoconstriction (including vasospastic conditions) and, together with the results of preclinical studies, suggested the particular importance of exploring the role of these agents in the treatment of hypertension and chronic heart failure.

The current evidence suggests that endothelin antagonists are very promising agents for the treatment of patients with chronic heart failure and already justifies the initiation of large scale clinical trials. The evidence of blood pressure reduction in hypertension remains to be confirmed and studies to determine additional benefits, for instance prevention or reversal of major additional risk factors such as cardiac hypertrophy, may well be needed before clinical development in this area will proceed. Importantly, endothelin-1 appears to be involved in atherogenesis. Evidence that endothelin antagonists could interrupt this process would markedly increase interest in, and expand the clinical potential for, this group of drugs. Finally, both ET_A selective and combined $ET_{A/B}$ receptor antagonists are currently in clinical development. Given the evidence that the predominant vascular effect of ET_B receptor stimulation appears to be endothelium dependent vasodilatation, it could be argued that ET_A selective receptor antagonists offer the most benefit. However, the effects of endothelin-1 extend beyond the vasculature and substantially more information is required before any firm predictions can be made as to which type of receptor antagonist is likely to offer most benefit to patients in specific cardiovascular indications. This is likely to be the major focus of much work in the immediate future and the publication of further clinical studies must be awaited with considerable interest.

References

1. Yanagisawa M, Kurihara H, Kimura S, Tomobe Y, Kobayashi M, Mitsui Y, Yazaki Y, Goto K, Masaki T (1988) A novel potent vasoconstrictor peptide produced by vascular endothelial cells. Nature 332:411–415
2. Webb DJ (1997) Physiological role of the endothelin system in human cardiovascular and renal haemodynamics. Curr Opin Nephrol Hypertens 1997; 6:69–73
3. McMahon EG, Palomo MA, Moore WM (1991)Phosphoramidon blocks the pressor activity of big endothelin[1–39] and lowers blood pressure in spontaneously hypertensive rats. J Cardiovasc Pharmacol 17 (Suppl 7):S29-S33
4. Clozel M, Breu V, Burri K, Cassal JM, Fischli W, Gray GA, Hirth G, Löffler BM, Müller M, Neidhart W, Ramuz H (1993)Pathophysiological role of endothelin revealed by the first orally active endothelin receptor antagonist. Nature 365:759–761
5. Véniant M, Clozel JP, Hess P, Clozel M (1994) Endothelin plays a role in the maintenance of blood pressure in normotensive guinea pigs. Life Sci 55:445–454
6. Kaasjager KAH, Shaw S, Koomans HA, Rabelink TJ (1997) Role of endothelin receptor subtypes in the systemic and renal responses to endothelin-1 in humans. J Am Soc Nephrol 8:32–39
7. Clarke JG, Benjamin N, Larkin SW, Webb DJ, Keogh BE, Davies GJ, Maseri A (1989) Endothelin is a potent long-lasting vasoconstrictor in men. Am J Physiol 257:H2033-2035
8. Wenzel RR, Noll G, Lüscher TF (1994) Endothelin receptor antagonists inhibit endothelin in human skin microcirculation. Hypertension 23:581–586
9. Haynes WG, Strachan FE, Webb DJ (1995) Endothelin ETA and ETB receptors cause vasoconstriction of human resistance and capacitance vessels in vivo. Circulation 92:357–363
10. Haynes WG, Webb DJ (1993) Endothelium dependent modulation of responses to endothelin-1 in human hand veins. Clin Sci 84:427–433
11. Cockcroft JR, Clarke JG, Webb DJ (1991) The effect of intra-arterial endothelin on resting blood flow and sympathetically mediated vasoconstriction in the forearm of man. Br J Clin Pharmacol 31:521–524
12. Haynes WG, Hand MF, Johnstone HA, Padfield PL, Webb DJ (1994) Direct and sympathetically mediated venoconstriction in essential hypertension: enhanced response to endothelin J Clin Invest 94:1359–1364
13. Yang Z, Richard V, von Segesser L et al (1990) Threshold concentrations of endothelin-1 potentiate contractions to norepinephrine and serotonin in human arteries: a new mechanism for vasospasm? Circulation 82:188–195
14. Haynes WG, Webb DJ (1994) Contribution of endogenous generation of endothelin-1 to basal vascular tone. Lancet 344:852–854
15. Love MP, Ferro CJ, Haynes WG, Webb DJ, McMurray JJ (1996) Selective or non-selective endothelin receptor blockade in chronic heart failure? Circulation 94 (Suppl 1):I74
16. Haynes WG, Ferro CJ, O'Kane KPJ, Somerville D, Lomax CC, Webb DJ (1996) Systemic endothelin receptor blockade decreases peripheral vascular resistance and blood pressure in man. Circulation 93:1860–1870
17. Plumpton C, Haynes WG, Webb DJ, Davenport AP (1995) Phosphoramidon inhibition of the in vivo conversion of big endothelin to endothelin-1 in the human forearm. Br J Pharmacol 116:1821–1828
18. Hemsen A, Ahlborg G, Ottoson-Seeberger A, Lundberg JM (1995) Metabolism of big endothelin-1 (1–38) and (22–38) in the human circulation in relation to production of endothelin (1–21). Regul Pept 55:287–297
19. Webb DJ (1995) The pharmacology of human blood vessels in vivo. J Vasc Res 32:2–15
20. Weber C, Schmitt R, Birnboeck H, et al (1996) Pharmacokinetics and pharmacodynamics of the endothelin-receptor antagonist bosentan in healthy human subjects. Clin Pharmacol Ther 60:124–137
21. Dubin D, Pratt RE, Dzau VJ (1989) ndothelin, a potent vasoconstrictor, is a vascular smooth muscle mitogen. J Vasc Med Biol 1:150–154
22. Battistini B, Chailler P, D'Orleans-Juste P, Briere N, Sirois P (1993) Growth regulatory properties of endothelins. Peptides 14:385–399
23. Tamamori M, Ito H, Adachi S, Akimoto H, Marumo F, Hiroe M (1996) Endothelin-3 induces hypertrophy of cardiac myocytes by the endogenous endothelin-1 mediated method. J Clin Invest 97:366–372
24. Clavell AL, Mattingly MM, Nir A, Aarhus LL, Heublein DM, Burnett, JC (1994a) Angiotensin converting enzyme inhibition modulates circulating and tissue endothelin activity in experimental heart failure. Circulation 90:I-452

25. Nishimura M, Takahashi H, Matsusawa M, Ikegaki I, Sakamoto M, Nakanishi T, Hirabayashi M (1991) Chronic intra-cerebroventricular infusions of endothelin elevate arterial pressure in rats. J. Hypertens 9:71–76

26. Rakugi H, Tabuchi Y, Nakamura M, Nagano M, Higashimori K, Mikami H, Ogihara T (1990) Endothelin activates the vascular renin-angiotensin system in rat mesenteric arteries. Biochem Int 21:867–872

27. Kohan DE, Padilla E (1993) Osmolar regulation of endothelin-1 production by rat inner medullary collecting duct. J Clin Invest 91:1235–40

28. Gellai M, DeWolf R, Pullen M et al (1994) istribution and functional role of renal ET receptor subtypes in normotensive and hypertensive rats. Kidney Int 46:1287–94

29. Panza JA, Quyyumi AA, Brush JE, Epstein SE (1990) Abnormal endothelium-dependent relaxation in patients with essential hypertension. N Engl J Med 323:22–27

30. Kohno M, Yasunari K, Murakawa KI et al (1990) Plasma immunoreactive endothelin in essential hypertension. Am J Med 88:614–618

31. Kohno M, Murakawa K, Horio T, Yokokawa K, Yasunari K, Fukui T, Takeda T (1991) Plasma immunoreactive endothelin-1 in experimental malignant hypertension. Hypertension 18:93–100

32. Davenport AP, Ashby MJ, Easton P et al (1990) A sensitive radioimmunoassay measuring endothelin-like immunoreactivity in human plasma: comparison of levels in patients with essential hypertension and normotensive control subjects. Clin Sci 78:261–264

33. Yokokawa K, Tahara H, Kohno M et al (1991) Hypertension associated with endothelin-secreting malignant haemangioendothelioma. Ann Intern Med 114:213–215

34. Tomobe Y, Miyauchi T, Saito A et al (1988) Effects of endothelin on the renal artery from spontaneously hypertensive and Wistar Kyoto rats. Eur J Pharmacol 152:373–374

35. Clozel M (1989) Endothelin sensitivity and receptor binding in the aorta of spontaneously hypertensive rats. J Hypertens 7:913–917

36. Schiffrin EL, Thibault G (1992) Blunted effects of endothelin upon small subcutaneous arteries of mild essential hypertensive patients. J Hypertens 10: 437–444

37. Schmitt R, Belz GG, Fell D, Lebmeier R, Prager C, Stahnke PL, Sittner WD, Karwoth A, Jones CR (1995) Effects of the novel endothelin receptor antagonist bosentan in hypertensive patients. Proceedings of the Seventh European Meeting on Hypertension, Milan

38. Ferro CJ, Haynes WG, Hand MF, Webb DJ (1996) Are the vascular endothelin and nitric oxide systems involved in the pathophysiology of essential hypertension? Eur J Clin Invest 26 Suppl.1:A51

39. Gray GA, Webb DJ (1996) The therapeutic potential of endothelin receptor antagonists in cardiovascular disease. Pharmacol Ther 72:109–48

40. Matin-Nizard F, Houssaini HS, Lestavel-Delattre S, Duriez P, Fruehart JC (1991) Modified low density lipoproteins activates macrophages to secrete ir-ET. FEBS Lett 293:127–130

41. Boulanger CM, Tanner FC, Bea ML, Hahn AWA, Werner A, Luscher TF (1992) Oxidised low density lipoprotein induces mRNA expression and release of endothelin from human and porcine endothelium. Circ Res 70:1(191–1197

42. Lerman A, Edwards BS, Hallett JW, Heublein DM, Sandberg SM, Burnett JC (1991) Circulating and tissue endothelin immunoreactivity in advanced atherosclerosis. N Engl J Med 325:997–1001

43. Dagassan PH, Breu V, Clozel M, Vogt P, Turina M, Kiowski W, Clozel JP (1996) Up-regulation of endothelin-B receptors in atherosclerotic human coronary arteries. J Cardiovasc Pharmacol 27:147–153

44. Zeiher AM, Ihling C, Pistorius K, Schachinger V, Schaefer H-E (1994) Increased tissue endothelin immunoreactivity in atherosclerotic lesions associated with acute coronary syndromes. Lancet 344:1405–1406

45. Tonnessen T, Giaid A, Saleh D, Naess PA, Yanagisawa M, Christensen G (1995) Increased in vivo expression of endothelin-1 by porcine cardiomyocytes subject to ischaemia. Circ Res 76:767–772

46. McMurdo L, Sessa WC, Thiermann C, Vane JR (1991) Ischaemia and reperfusion injury potentiates the vasoconstrictor effects of endothelin-1 in the isolated perfused heart of the rat. Br J Pharmacol 104:343P

47. Liu J, Chen R, Casley DJ, Nayler WN (1990) Ischaemia and reperfusion increase 125-I labelled endothelin-1binding in rat cardiac membranes. Am J Physiol 258:H828–835

48. Ray SG, McMurray JJ, Morton JJ, Dargie HJ (1992) Circulating endothelin in acute ischaemic syndromes. Br Heart J 67:383–386

49. Wieczorek I, Haynes WG, Webb DJ, Ludlam CA, Fox KA (1994) Raised plasma endothelin in unstable angina and non-Q wave myocardial infarct ion: relation to cardiovascular outcome. Br Heart J 72:436–441
50. Yasuda M, Kohno M, Tahara A et al (1990) Circulating immunoreactive endothelin in ischemic heart disease. Am Heart J 119:801–806
51. Lechleitner P, Genser N, Mair J, Maier J, Artner-Dworzak E, Dienstl F, Puschendorf B (1992) Endothelin-1 in patients with complicated and uncomplicated myocardial infarction. Clin Investig 70:1070–1072
52. Omland T, Lie RT, Aakvaag A (1994) Plasma endothelin determination as a prognostic indicator of 1-year mortality after acute myocardial infarction. Circulation 89:1573–1579
53. Dargie HJ, McMurray JJV (1994) Diagnosis and management of heart failure. Br Med J 308:321–328
54. Swedberg K, Eneroth P, Kjekshus J (1990) Hormones regulating cardiovascular function in patients with severe congestive heart failure and their relation to mortality. Circulation 82:1730–1736
55. Pacher R, Bergler-Klein J, Globits S, Teufelsbauer H, Schuller M, Krauter A, Ogris E, Rodler S, Wutte M, Hartter E (1993) Plasma big endothelin-1 concentrations in congestive heart failure patients with or without systemic hypertension. Am J Cardiol 71:1293–1299
56. Cody RJ, Haas GJ, Binkley PF, Capers Q, Kelley R (1992) Plasma endothelin correlates with the extent of pulmonary hypertension in patients with chronic congestive heart failure. Circulation 85:504–509
57. Krum H, Goldsmith R, Wilshire-Clement M, Miller M, Packer M (1992) Importance of endothelin in the exercise intolerance of heart failure in humans. Mayo Clin Proc 67:719–724
58. Galatius-Jensen S, Wroblewski H, Emmeleuth C, Bie P, Haunso S, Kastrup J (1994) Plasma endothelin-1 in chronic heart failure: a predictor of cardiac death? Circulation 90: 379.
59. Kiowski W, Sütsch G, Hunziker P, Müller P, Kim J, Oechslin E, et al (1995) Evidence for endothelin-1 mediated vasoconstriction in severe chronic heart failure. Lancet 346:732–736
60. Pacher R, Stanek B, Hülsmann M, Koller-Strametz J, Berger R, Schuller M et al (1996) Prognostic impact of big endothelin-1 plasma concentrations compared with invasive hemodynamic evaluation in severe heart failure. J Am Coll Cardiol 27:633–641
61. Yoshida H, Nakamura M (19*) Inhibition by angiotensin converting enzyme inhibitors of endothelin secretion from cultured human endothelial cells. Life Sci 22:195–200
62. Abernethy DR, Laurie N, Andrawis NS (1995) Local angiotensin-converting enzyme inhibition blunts endothelin-1 induced increase in forearm vascular resistance. Clin Pharmacol Ther 58:328–334
63. Davidson NC, Coutie WJ, Webb DJ, Struthers AD (1996) Neurohormonal and renal differences between low-dose and high-dose ACE-inhibitor treatment in patients with chronic heart failure. Heart (in press)
64. Townend J, Doran J, Jones S, Davies M (1994) Effect of angiotensin converting enzyme inhibition on plasma endothelin in congestive heart failure. Int J Cardiol 43:299–304
65. Krum H, Gu A, Wilshire Clement M, Sackner Bernstein J, Goldsmith R (1996) Changes in plasma endothelin-1 levels reflect clinical response to beta blockade in chronic heart failure. Am Heart J 131:337–341
66. Love MP, Haynes WG, Gray GA, Webb DJ, McMurray JJV (1996) Vasodilator effects of endothelin-converting enzyme inhibition and endothelin ETA receptor blockade in chronic heart failure patients treated with ACE inhibitors. Circulation 94:2131–2137
67. Cannan CR, Burnett JC Jr, Lerman A (1996) Enhanced coronary vasoconstriction to endothelin-B-receptor activation in experimental congestive heart failure. Circulation 93:646–651
68. Pfeffer MA, Pfeffer JM, Steinberg C, Finn P (1985) Survival after an experimental myocardial infarction: beneficial effects of long-term therapy with captopril. Circulation 72:406–412
69. Sakai S, Miyauchi T, Kobayashi M, Yamaguchi I, Goto K, Sugishita Y (1996) Inhibition of myocardial endothelin pathway improves long-term survival in heart failure. Nature 384:353–355

Subject Index

Printing: Saladruck, Berlin
Binding: Buchbinderei Lüderitz & Bauer, Berlin